Industrial and Hazardous Wastes Treatment

Industrial and Hazardous Wastes Treatment

Zander Ellis

SYRAWOOD
PUBLISHING HOUSE

New York

Published by Syrawood Publishing House,
750 Third Avenue, 9th Floor,
New York, NY 10017, USA
www.syrawoodpublishinghouse.com

Industrial and Hazardous Wastes Treatment
Zander Ellis

International Standard Book Number: 978-1-68286-824-9 (Hardback)

Cataloging-in-Publication Data

Industrial and hazardous wastes treatment / Zander Ellis.
    p. cm.
Includes bibliographical references and index.
ISBN 978-1-68286-824-9
1. Factory and trade waste. 2. Hazardous wastes. 3. Hazardous wastes--Purification.
4. Refuse and refuse disposal. 5. Salvage (Waste, etc.) I. Ellis, Zander.
TD897 .I53 2019
628.4--dc23

TABLE OF CONTENTS

Preface VII

Chapter 1 **Introduction to Industrial and Hazardous Waste** 1
- Industrial Waste 1
- Hazardous Waste 5

Chapter 2 **Characteristics of Hazardous Waste** 23
- Ignitability 23
- Corrosivity 25
- Reactivity 27
- Toxicity 29

Chapter 3 **Pharmaceutical Waste Management** 32
- Pharmaceutical Waste 32
- Types of Medical Waste 35
- Storage of Medical Waste 73
- Treatment of Medical Waste 79
- Disposal of Medical Waste 83

Chapter 4 **Oil Refinery Waste Treatment** 90
- Oil Refinery Waste 90
- Rehabilitation of Storage Tanks 100
- Treatment of Semi-processed Oil Refinery Products 101
- Remediation of Hydrocarbon Contaminated Soils 103

Chapter 5 **Textile Waste Treatment** 113
- Textile Waste 113
- Silk Waste 121
- Textile Recycling 122
- Textile Wastewater Treatment 126

Chapter 6 **Other Industrial Wastes and their Treatment** 130
- Photographic Waste Treatment 130
- Pesticide Waste Treatment 138
- Pulp and Paper Mill Wastes Treatment 158

- Rubber Industry Wastes Treatment.. 174
- Timber Industry Wastes Treatment.. 182
- Soft Drink Waste Treatment 183
- Bakery Waste Treatment 197
- Power Industry Wastes Treatment 205

Permissions

Index

PREFACE

Hazardous wastes refer to substances that have a harmful or deteriorative effect on public health and the environment. This may be due to the toxicity, flammability, corrosivity, reactivity or radioactivity of the hazardous wastes. They can be recycled to produce new products, or destroyed by high temperature incineration or pyrolysis. Wastes can also be treated using cement based solidification and stabilization. Industrial wastes are any material rendered useless during an industrial process and may include chemical solvents, sludge, metals, pigments, paints and radioactive wastes. Such toxic wastes require specialized treatments. These can be classified under chemical waste, industrial solid waste, toxic waste and municipal solid waste. This book aims to shed light on some of the unexplored aspects of industrial and hazardous waste. It also elucidates the modern techniques of waste treatment and their applications. In this book, constant effort has been made to make the understanding of the difficult concepts of this field as easy and informative as possible, for the readers.

A foreword of all Chapters of the book is provided below:

Chapter 1- Industrial waste is the waste that is produced due to industrial activity. These wastes generally include materials that are considered undesirable during manufacturing processes in industries, factories and mining operations. Hazardous waste is the waste that can pose a threat to the environment or public health, such as discarded chemical products, radioactive wastes, etc. This chapter has been carefully written to provide an extensive understanding of industrial and hazardous waste.;

Chapter 2- Hazardous wastes are present in the state of solid, liquid or gas. Some of the characteristics of hazardous wastes are reactivity, ignitability, toxicity and corrosivity. This chapter discusses in detail the different properties of hazardous waste for a comprehensive understanding.;

Chapter 3- Biomedical waste is the waste, which contains infectious materials. Any waste of laboratory or medical origin such as microbial cultures, blood, body parts, human or animal tissue, body fluids, etc. can be termed biomedical or pharmaceutical waste. The management of such waste is vital for human health and safety. This chapter closely examines the different types of medical waste, its storage, treatment and disposal.;

Chapter 4- Oil refinery is an industrial plant concerned with the refinement of crude oil into useful products like gasoline, diesel fuel, kerosene, LPG, etc. It produces a considerable amount of industrial waste. All the diverse aspects of oil refinery waste treatment such as treatment of semi-processed oil refinery products, remediation of hydrocarbon contaminated soils, rehabilitation of storage tanks, etc. have been covered extensively in this chapter.;

Chapter 5- The textile industry is concerned with the production, design and distribution of garments, cloth and yarn. The manufacturing of textiles results in large amounts of hazardous waste. The topics elaborated in this chapter such as silk waste, textile recycling, textile waste and textile wastewater treatment will provide a detailed understanding of waste, generated by the textile industry and its treatment.;

Chapter 6- Industries produce a lot of waste such as chemical solvents, sludge, pigments, ash, metal, radioactive wastes etc. These require specialized treatment systems for safe disposal, which are different depending on whether the waste is solid, liquid or radioactive. The aim of this chapter is explore the common industrial waste treatment methods, for photographic waste treatment, pesticide waste treatment, rubber industry wastes treatment, pulp and paper mill wastes treatment, etc.

I would like to thank the entire editorial team who made sincere efforts for this book and my family who supported me in my efforts of working on this book. I take this opportunity to thank all those who have been a guiding force throughout my life.

Zander Ellis

Chapter 1

Introduction to Industrial and Hazardous Waste

Industrial waste is the waste that is produced due to industrial activity. These wastes generally include materials that are considered undesirable during manufacturing processes in industries, factories and mining operations. Hazardous waste is the waste that can pose a threat to the environment or public health, such as discarded chemical products, radioactive wastes, etc. This chapter has been carefully written to provide an extensive understanding of industrial and hazardous waste.

Industrial Waste

Our Environment is being polluted every day. One of the main reasons behind this pollution is the industrial waste. Factories are dumping their waste anywhere thy can, and it is polluting our environment. Since the industrial revolution, small and traditional trades have been replaced by large industrial factories. Industrialization has been the result of rapid developments in technology and the increasing demands in our daily lives. These industries are essential to the progress of human civilization. There is almost no alternative to industrialization for meeting the high worldwide demand for goods. But the main thing is that most industrial factories are not eco-friendly at all. These factories produce a lot of waste every day. They are dumping their chemical and other waste wherever they want and polluting the air, water, and land. It is causing the soil pollution and climate change which is dangerous for earth's biodiversity.

Wastes discharged from factories are polluting our rivers and other water bodies with chemicals. Dyes from garments factories and waste from tanneries end up in the rivers. Due to these toxic chemicals, the water has become nothing less than dye. The major victims of water pollution are amphibians and dolphins. A lot of amphibians that hunt insects in our farmlands are dying off. Just because of this, several species of fish and aquatic plants are endangered today. A lot of important plants are dying off due to toxic chemicals in the water. This harms the animals dependent on these plants. Fishes lose their breeding grounds and harm the overall ecosystem. This toxic water is also added on our farmlands by irrigation and because of this, the toxic chemicals are making their way into our food and cause diseases. Sometimes, forests are cut down to make space for these factories. This has caused habitat loss for many animals.

Polythene, plastics, and electronics also result in environmental pollution. Polythene and plastics are not biodegradable and finally, they end up collecting on river beds. The uppermost layer of the soil is used for either farming or for construction and this layer of soil is getting destroyed every day due to hard waste collecting on the soil. Because of this, a lot of tall buildings are now safety hazards as the top soil is being harmed by solid wastes. The beneficial invertebrates that dwell in the soil are also lost and the proper bonding of soil is damaged by it which results in the reduced production of harvest. The main factor behind air pollution is smoke from factories. Toxic substances like Carbon-monoxide are released into the air from this smoke. Brick kilns are responsible for both deforestation and air pollution from the smoke. Humans and other animals alike are under heavy threat because of air pollution. Increased industrialization has led to increased unplanned urbanization. One of the largest factors exacerbating global warming is industrial factories. So, the climate change effect is fast forwarding and floods, cyclones, tsunamis etc. have all increased in number and strength due to climate change. Coastal and oceanic biodiversity are being destroyed as a result. Some species that have been greatly harmed are sea turtles, corals, snails, oysters, crabs and different types of birds. This in turn gravely affects the lives of coastal people. Unsustainable industrialization and pollution are having long term effects on our environment. So, before creating any factories, the environmental factors need to be considered. The waste management process at every factory needs to be improved and used. Some industrial wastage treatment plants need to be established near factories and other production places so that the water bodies can be protected. In these plants, the wastes from factories are purified stepwise. Hard wastes like plastics and polyethylene need to be recycled. Along with this, the methods of transportation and electricity production need to be environmentally sustainable. Sustainable development will be beneficial to both humans and the environment.

Today, there are still a lot of factories that don't use the treatment plants and recycling methods. When someone sets up a new industry, they are only concerned about the production not the environment and these entrepreneurs probably don't even have the necessary knowledge. So, the government should have some sort of program that can encourage and educate them about the effects of industrial biodegradable waste. We need industries for modern life and it is obvious that they will produce waste. We just need to increase awareness about the waste treatment. There are many private organizations working in the field of waste treatment and the industries need to outsource the waste material to them. These contractors will collect all the liquid and solid waste and treat it. Then the business can focus on only their products without worrying about treating the waste. There is a need to establish central treatment plants instituted by the government for each zone. Thermal power plants should also be established so that the large cooker can be replaced in the factories. This will help in eliminating smoke from burning coal. Afforestation is also another measure as if there is a heavily forested area, there will be less wind speed and the dust particle won't be able to travel long distance with the wind. Another measure is to impose heavy penalties on the industries not following the pollution control rules. We just need to realize that it is our duty to keep the pollution at its lowest level.

It is obvious that with increasing industrialization the pollution will also increase but still, we have to follow the measures to survive alongside. We can also take some small steps on our behalf like reducing the amount of garbage that goes into the dust, reusing & recycling the items so that a number of new items can be decreased, using less paper, not using vehicles all the time and taking a walk instead. The factories should use chimneys, treat the wastes and disposals, degrade toxicants & poisonous materials, adopt green accounting system, use less of agrochemicals and fulfill the corporate social responsibility. These all measures can help us prevent industrial pollution, at least, to some extent.

We have millions of factories, mills, industries, mining plants, etc. around the world. These industries use raw materials to produce finished goods for consumers. But in the manufacturing process, there are materials which are rendered useless. They constitute the industrial waste. Some examples of industrial wastes are metals, paints, sandpaper, slag, ash, radioactive wastes, etc.

Types of Industrial Waste

Industrial waste can be categorized into biodegradable and non-biodegradable:

- Biodegradable: Those industrial wastes which can be decomposed into the non-poisonous matter by the action of certain microorganisms are the biodegradable wastes. They are even comparable to house wastes. These kinds of waste are generated from food processing industries, dairy, textile mills, slaughterhouses, etc. Some examples are paper, leather, wool, animal bones, wheat, etc. They are not toxic in nature and they do not require special treatment either. Their treatment processes include combustion, composting, gasification, bio-methanation, etc.

- Non-biodegradable: Those industrial wastes which cannot be decomposed into non-poisonous substances are the non-biodegradable wastes. Examples are plastics, fly ash, synthetic fibers, gypsum, silver foil, glass objects, radioactive wastes, etc. They are generated by iron and steel plants, fertilizer industries, chemical, drugs, and dyes industries. It is estimated that about 10 to 15 percentage of the total industrial wastes are non-biodegradable and hazardous, and the rate of increase in this category of waste is only increasing every year. These wastes cannot be broken down easily and made less harmful. Hence, they pollute the environment and cause threat to living organisms. They accumulate in the environment and enter the bodies of animals and plants causing diseases. However, with the advancement in technology, several disposals, and reuse methods have been developed. Wastes from one industry are being treated and utilized in another industry. For example, the cement industry uses the slag and fly ash generated as waste by steel industries. Landfill and incineration are other methods which are being resorted to, for the treated of hazardous wastes.

- With so many different categories of industrial waste, it's important to understand what materials can be reused and recycled, and how to properly manage and reduce waste disposal. Here are some facts about the different categories of industrial waste.

Chemical Waste

Chemical waste is typically generated by factories, processing centers, warehouses, and plants. This waste may include harmful or dangerous chemicals and chemical residue, and waste disposal must adhere to careful guidelines. These guidelines are instituted and regulated by various government and environmental agencies, such as the Environmental Protection Agency, and the Occupational Safety and Health Administration. There are generally fines associated with non-compliance. Chemical waste must be segregated on-site, and waste disposal may need to be handled by a specialist to ensure compliance with health, safety, and legal requirements.

Solid Waste

In industrial services, solid waste includes a variety of different materials, including paper, cardboard, plastics, packaging materials, wood, and scrap metal. Some of these materials can be reused and recycled by a recycling center. If you don't have a comprehensive waste management plan that includes recycling, your waste disposal is not going to be as cost-effective or environmentally friendly as it could be. A recycling center can process the majority of industrial solid waste, effectively reducing your waste disposal costs.

Toxic and Hazardous Waste

Toxic and hazardous waste is comprised of materials that can cause serious health and safety problems if waste disposal is not handled correctly. This type of waste typically includes dangerous byproducts materials generated by factories, farms, construction sites, laboratories, garages, hospitals, and certain production and manufacturing plants. The EPA and state departments regulate toxic and hazardous waste disposal. This waste disposal is only legal at special designated facilities around the country.

Hazardous Waste

Any solid waste, other than radioactive wastes, which by reasons of physical and/ or chemical or reactive or toxic, explosive, corrosive or other characteristics causing danger or likely to cause danger to health or environment whether alone or when coming in contact with other waste or environment.

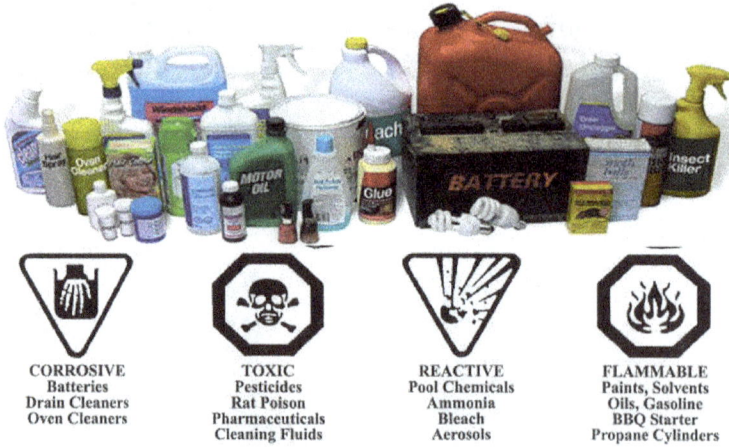

CORROSIVE	TOXIC	REACTIVE	FLAMMABLE
Batteries	Pesticides	Pool Chemicals	Paints, Solvents
Drain Cleaners	Rat Poison	Ammonia	Oils, Gasoline
Oven Cleaners	Pharmaceuticals	Bleach	BBQ Starter
	Cleaning Fluids	Aerosols	Propane Cylinders

Classification of Hazardous Waste

It is important to have a clear understanding of the nature of materials involved to discuss the techniques for its management. The understanding is not limited to the physical state or composition but also to a complete scheme of characterization that includes all information necessary for disposal or recycling after treatment.

A means of characterizing the waste is needed by the producer, disposer and the recycle. Practically, producer often does not have sufficient information, support, motivation or incentives to carry out treatment and/or disposal of hazardous waste and hence greater responsibility lies with the disposal contractor, if appointed for secured landfill out-side the company premises.

In order to help the producer, disposer and recycle in the identification of hazardous waste, United State Environment Protection Agency describe the testing of solid wastes for various hazard parameters such as corrosively, reactivity, ignitability as well as toxicity to designate the waste as hazardous (table).

Table: Parameters for Hazard Potential

Parameter	Defining Characteristics	Example
Corrosivity	Waste which have pH < 2.0 or > 12.0 or which corrodes steel at a rate greater than 6.35 mm per year at 55°C.	Acidic waste, spent pickle liquor.
Reactivity	Wastes, which are unstable and spontaneously react with water or air, generate toxic gases and explode due to shock or heat.	Water from TNT operation and used Cyanide solvents.
Ignitability	Wastes which spontaneously ignite in dry or moist air at or below 60°C.	Waste oils, used solvents.
Toxicity	Wastes which release toxic materials on leaching in sufficient amounts to pose a substantial hazard to human health or environment as measured by the Toxicity Characteristics Leaching Procedure (TCLP)	Metals bearing wastes.

Management of Hazardous Waste

Any sort of hazardous waste finds its management in either of the following ways:

1. Waste minimization.

2. Waste treatment.

3. Waste disposal.

Waste Minimization, Reuse/Recovery

The green productivity concept always supports and encourages waste minimization for Cleaner production. The fact that reduction is better than management is a vital factor in making any strategy at the time of commissioning of any industrial project. More often this aspect is overlooked by other measures of material productivity like, manpower deployment, labor productivity and energy consumption and project out-put.

In-fact, waste reduction is a way of improving profitability and competitiveness. It not only minimize the need of abiding by pollution control regulations that force to spend more and more money for rather smaller increments of environmental protection but also reduces risks associated with the generation of such wastes. Waste -reuse or recycle is usually the step before pollution control but the economic limitations have to be taken into account.

Following are the various approaches to waste minimization and its reuse:

1. Alternative usage of waste products.

2. Modifying production process.

3. Altering primary source of waste generator by improving process technology and equipment.

4. Improving plant operations such as better house-keeping, improved material handling, and equipment maintenance, automating process equipment, better monitoring and improved waste tracking.

5. Optimizing process conditions.

6. Introducing substitute raw materials which have a lesser potential of generating hazardous waste.

7. Redesigning or reformulating end products.

8. Segregating usable wastes and waste.

9. Segregating waste and hazardous waste.

10.   Transferring the waste to another industry that can utilize it.

11.   Reprocessing waste to recover energy or material.

12.   Recycling potential waste or portion of it to the generator site.

Table: Recovery Options for Management of Hazardous Waste

S. No.	Type of Waste	Treatment
1.	Solvent	Distillation, filtration, evaporation, centrifugation and stripping
2.	Oil	Oil re-refining, distillation
3.	Used acid	Acid regeneration, Filtration through Ion exchange
4.	Metals	Ion exchange, electrodialysis, reverse osmosis, membrane filtration, adsorption, reverse plating, solvent stripping and precipitation
5.	Fuel	Fuels blending with waste oils, solvents and still bottoms with high BTU fuels or coal

Waste Treatment

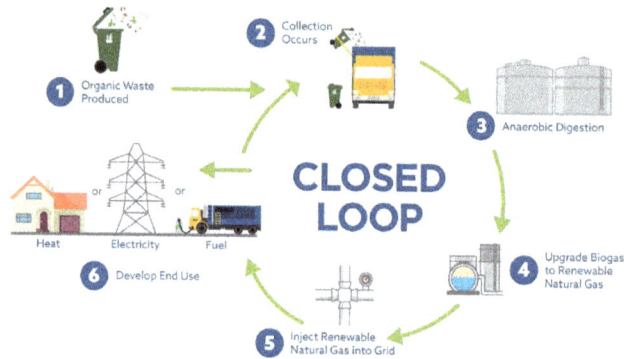

All the waste products whether from manufacturing process or treatment facility must be treated for the impurities hazardous to the nature to render them harmless to the environment.

The various treatment procedures can be classified as:

1.   Physical.

2.   Chemical.

3.   Biological.

4.   Thermal.

Physical Treatment

Physical treatment of hazardous waste includes a number of separation processes commonly used in industry. It is of first importance where waste containing liquids and solids are separated to reduce cost.

Chemical Treatment

These procedures involve the use of chemical reactions with the help of various chemicals to convert hazardous waste into less hazardous substances. The chemical treatment produces useful by- products and some-times residual effluent that are environmentally acceptable. Chemical reactions either reduce the volume of the waste or convert the wastes to a less hazardous form.

Biological Treatment

Biological treatment is an effective, efficient and cost- effective way to treat7remove hazardous substances from wastewater through biological agents. Hazardous waste materials are toxic to some of the microorganism. But a substance, which is toxic to one group of organism, may act as valuable source of food for another group.

Bio-treatment is required in ideal conditions for better growth of bio-agents and hence is a limitation factor also. Biological systems can lower the cost of downstream processes by reducing organic load if they are supplemented by other physical or chemical treatment steps.

Thermal Treatment

Components of most hazardous wastes are carbon, hydrogen, oxygen, halogens, sulphur, nitrogen and heavy metals. Due consideration should be given to these constituents while applying incineration technology to the thermal destruction of hazardous waste.

In incineration, in general, waste is destroyed or reduced to CO_2, H_2O and other inorganic substances and these substances are harmless. The only limitation with this treatment process is generation of effluent or emission which is rather secondary pollution. Various treatment options for hazardous wastes are summarized in Table.

Table: Treatment Options for Management of Hazardous Waste

S. No.	Type of Waste	Treatment
1.	Cyanide bearing.	Detoxification/biological.
2.	Heavy metal bearing.	Sludge conditioning, dewatering, metal recovery.
3.	Non-halogenated or halogenated hydrocarbon including solvents.	Thermal treatment after solvent recovery.
4.	From paint, pigment, glue, varnish, ink industry.	Thermal/biological.
5.	From dyes and intermediate containing inorganics.	Depending on type of waste.
6.	From dyes and intermediate containing organics.	Thermal.
7.	Waste oil and emulsions.	Physico-chemical and biological for oil separation and thermal treatment for sludge.
8.	Tarry waste and residues from refining, cracking.	Thermal.
9.	Sludge from treatment plants.	Sludge conditioning and dewatering.
10.	Phenols.	Detoxification.
11.	Asbestos.	Preparing for landfill.
12.	Wastes and residues from pesticides.	Thermal, biological treatment and solidification.
13.	Acid/alkali slurry.	Physico-chemical treatment, sludge for conditioning and dewatering.
14.	Off-spec. products.	Physico-chemical, biological and thermal.
15.	Discarded containers and liners.	Thermal for liners and landfill for containers.

Various physical, chemical and biological treatment processes are depicted below:

PHYSICAL TREATMENT PROCESS

- Screening
- Centrifugation
- Filtration
- Evaporation & distillation
- Stripping
- Sedimentation & clarification
- Flotation
- Sorption
- Reverse osmosis

CHEMICAL TREATMENT PROCESS

- Neutralization
- Solubility
- Caogulation and flocculation
- Colour removal
- Precipitation
- Oxidation & reduction
- Disinfection

BIOLOGICAL TREATMENT PROCESS

- Bioremediation
- Aerobic/anaerobic reactor
- Metals uptake through plant species
- Wetland technology
- Composting method
- Bacterial culture

Waste Disposal

This is an ultimate option with every industry. Depending upon the characteristics of the wastes, two types of disposal methods can be used for hazardous wastes.

The predominant method for hazardous wastes disposal after treatment and reuse are:

1. Landfill.

2. Incineration.

Landfill

Landfills are necessary because one cannot totally eliminate generation of hazardous waste and treatment technologies produce residues.

Landfills Involve

1. Low permeability soil linear and/or synthetic linear to prevent seepage of leachate to underground strata.

2. Drainage layer for leachate collection with leak detection system.

3. Diversion of surface run-off to avoid further pollution.

4. Impermeable cover.

5. Groundwater monitoring wells.

Issues with Secure Landfills

Landfill options are highly risky options if not taken care. There is always threat of subsurface pollution.

The other considerations are as follows:

i. Only solid wastes are landfilled. Liquid wastes are solidified before being placed in a secure landfill.

ii. Landfill slope stability-subsidence of the cells is taken into account.

iii. Proper segregation and tracking of wastes is necessary to know where the waste has been placed in the compartment.

iv. Record keeping is must.

v. Maintenance of landfill cover integrity and monitoring of ground water. Various disposal options for hazardous wastes are summarized in Table.

Table: Disposal Options for Management of Hazardous Waste

S. No.	Type of Waste	Disposal
1.	Cyanide bearing	Sludge for land-fill
2.	Heavy metal bearing.	Land-fill.
3.	Non-halogenated or halogenated hydrocarbon including solvents.	Thermal treatment and ash for land-fill.
4.	From paint, pigment, glue, varnish, ink industry.	Thermal treatment and ash for land-fill.
5.	From dyes and intermediate containing inorganics.	Land-fill.
6.	From dyes and intermediate containing organics.	Thermal treatment and ash for land-fill.
7.	Waste oil and emulsions.	Sludge for incineration and then ash for land-fill.
8.	Tarry waste and residues from refining, cracking.	Thermal treatment and ash for land-fill.
9.	Sludge from treatment plants.	Land-fill.
10.	Phenols.	Sludge for land-fill.
11.	Asbestos.	Land-fill.
12.	Wastes and residues from pesticides.	Thermal treatment, land-fill.
13.	Acid/alkali slurry.	Sludge for land-fill.
14.	Off-spec. products.	Ash and sludge for land-fill.
15.	Discarded containers and liners.	Containers for land-fill and liners for incineration.

Household Hazardous Waste

Discarded waste in our households may contain hazardous and toxic chemicals. The typical Indian household uses many products that may contain various substances that can be categorized as hazardous. These include insecticides, pesticides and fungicides; wood preservatives; broken CFLs, tube lights; paints, thinners, stains and varnishes; adhesives and glues; medicines, cosmetics, nail polish and removers; batteries; various cleaners and polishes; and variety of electrical and electronic items.

This fact sheet intends to describe some of the potential dangers associated with Household Hazardous Wastes (HHW), as well as to provide information about how to properly manage and prevent the generation of HHW.

A 2006 survey amongst selected Delhi residents conducted by Toxics Link revealed that every household has some or other kind of household hazardous product that are mostly thrown away with the municipal waste. The table at the bottom of the page provides the number of households with improper hazardous household waste disposal. This can include pouring them down the drain, on the ground, into sewers, and most commonly in Indian context putting them out with the household general waste bin.

The dangers of such disposal methods might not be immediately obvious, but has every possibility to pollute the environment and pose a threat to human health. Some HHWs can cause physical injuries to sanitation workers, may contaminate our drinking water if poured down drains, toilets, or on the ground, and create hazards for children and pets.

Amount of hazardous waste disposed by households

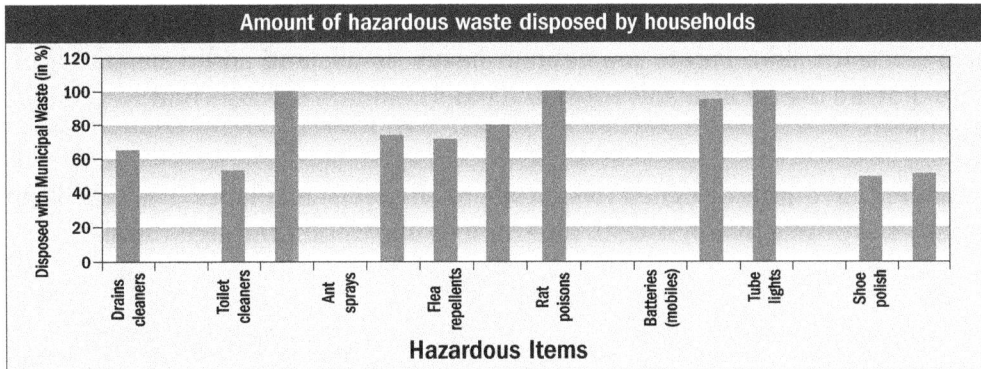

The household waste can be described as:

- Products are considered as hazardous waste owing to their potential to harm the environment when disposed of incorrectly.

- Household hazardous waste is any product with the word danger (most hazardous) or warning or   caution (less hazardous) on its label.

- A hazardous product will also have at least one of the following properties on its label:

  ○ Toxic

  ○ Flammable

  ○ Corrosive

  ○ Reactive

Risks of Household Hazardous Waste

Improper storage of chemicals in your household can turn out to be harmful to children or pets and be a fire hazard. Chemicals poured down the drain pollute our drinking water and can contaminate septic tanks and waste water treatment facilities. When thrown in the trash, some household hazardous waste can harm sanitation workers. In a line, whatever we do with our household hazardous wastes will affect everyone.

Household cleaners may contain solvents that are hazardous to breathe and can get into the body and skin. Many are irritants and can react with ammonia to create a toxic gas. Polishes usually contain petroleum distillates, which catch fire easily and can be hazardous to inhale.

Automotive products contain pollutants that are poisonous and catch fire. A small amount of oil, if disposed of improperly, can contaminate large quantities of drinking water.

Car batteries contain lead and sulphuric acid. The lead can contaminate the water and the acid can burn skin. Paints can contain heavy metals and additives, like lead and mercury that are toxic. Oil or solvent-based paints contain solvents that can be harmful to your lungs.

Pesticides are chemicals designed to kill rodents, insects, and plants. They can injure or potentially kill people by inhalation, ingestion or absorption through the skin.

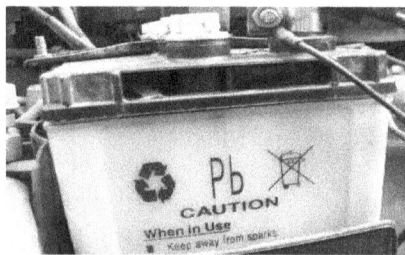

Automobile batteries should never be mixed with garbage. Always give those back to battery recyclers.

Items that may be considered as a household hazardous waste:

- Paint products,

- Beauty products, such as nail polish remover,

- Mercury-containing items such as thermometers,

- Empty plastic cans of mosquito repellants,
- Used up cans of termite, insect and cockroach repellent,
- Discarded Medicines,
- Household cleaners – Toilet Bowl cleaner, floor cleaner, other disinfectants.

Types of Household Hazardous Waste

Cleaning Products:

- Oven cleaners
- Drain cleaners
- Wood and metal cleaners and polishes
- Toilet cleaners
- Tile, shower cleaners
- Bleach (laundry)

Indoor Pesticides:

- Ant sprays and baits
- Cockroach sprays and baits
- Flea repellents and shampoos
- Bug sprays
- Houseplant insecticides
- Mosquito coils
- Moth repellents

Automotive Products:

- Motor oil
- Fuel additives
- Air conditioning refrigerants
- Starter fluids
- Automotive batteries
- Transmission and brake fluid
- Antifreeze

Painting Supplies:

- Adhesives and glues
- Furniture strippers
- Oil or enamel based paint
- Latex
- or water based paint
- Stains and finishes
- Paint thinners and turpentine
- Paint strippers and removers
- Fixatives and other solvents
- Herbicides
- Insecticides
- Fungicides/wood preservatives
- Batteries
- Computers and electronics
- Mercury thermostats or thermometers
- Fluorescent light bulbs
- Discarded PVC Toys

Other Flammable Products:

- Compressed gas cylinders
- Kerosene
- Diesel fuel
- Gas/oil mix
- Lighter fluids
- Shoe

Women and HHWs

HHWs have a huge impact on women's health primarily because in the Indian context till date, women in the families handle the bulk household chores. Starting from cleaners to compressed gas cylinders or distorted freezer racks, women are in contact with all. In addition to these women are the primary consumers of most of the beauty products like nail polish, removers and shampoos that in process turn out to be HHWs. Therefore with lack of proper disposal and waste management systems in our society women become extremely vulnerable towards health risks pertaining to HHWs.

Measures to be Taken

1. Read the Label and Refuse to Buy

Reading the label on a product can pro-vide a great deal of information including its ingredients, use, proper storage, and occasionally even disposal instructions. If the product contains hazardous ingredients, certain key words (amongst the list of unrecognizable chemical ingredients) can guide the purchaser to contents that can pose an environmental threat if not managed properly:

- For pesticides: Danger means highly toxic; Warning means moderately toxic; and Caution means slightly toxic.

- For general household products:

POISON means highly toxic; Danger means extremely flammable, corrosive, or highly toxic; and Warning or Caution means less toxic, but still dangerous.

Identify hazardous products by checking the label for key words like:

- DANGER: highly toxic, flammable or corrosive

- POISON: highly toxic

- WARNING: moderately toxic

- CAUTION: slightly toxic

- TOXIC: may cause death, cancer, or birth defects if swallowed, inhaled or comes in contact with skin

- FLAMMABLE: will burn or explode if exposed to certain temperatures or conditions

- CORROSIVE: can cause serious damage when it come in contact with skin

Other words on the label that indicate whether a product is hazardous include the following:

- Wear gloves" is a sign of corrosive or dermal toxic substances.

- "Do not store near heat or open flame" suggests ignitability or flammability.

- "Do not store near..." indicates reactive qualities of the material.

- "Use only in well ventilated room" is used for toxic fumes and reactive chemicals.

Use alternative products or homemade remedies and cleaners. If you have domestic cleaning services or painting and repair contractors, request that they only use non-hazardous products, for example latex paint without lead and other heavy metals. To learn about the use of alternative methods or products without hazardous ingredients refer to:

Toxics Link Fact sheet No. 2, "Toxics Free Home", August 1998. Toxics Link Fact sheet No. 9, "Mosquito Bites", July 2000.

Following is the list of some items in our households that contain PVC. This list is not complete but it is comprehensive and provides a descent understanding:

- Cleaning product containers
- Clothes racks (covers metal to prevent rusting)
- Vinyl file folder covers
- Fake decorative trees
- Credit and debit cards
- Imitation leather furniture
- Mattress covers
- Pet care product containers
- Photo album sheets
- Self-adhesive labels and stickers
- Shelving
- Strollers
- Shower curtains

- Textiles
- Toys
- Appliance casings
- Dishwasher, refrigerator and freezer racks
- Drinking straws
- Food containers
- Food wrap
- Plastic utensils
- Table covers etc.
- Many more.

2. Store and use Properly

In case you purchase hazardous products buy only what you need and use it up completely. When using, read and follow the instructions on the label. Wear the proper protection and use the right amount. Always re-seal the unused portion between usages. Store the products safely. Never store hazardous products in food containers. Keep them in their original containers and never remove labels. Secure hazardous products away from children and pets. Do not mix chemicals together; they might react, ignite, or explode.

3. Reuse and Recycle

Share any leftover products with neighbours, family, friends, or organizations that can use them. Any products that cannot be used or given away should be properly disposed.

Burnt Batteries found in waste dump

4. Disposal

If you have no other option, follow these instructions for proper disposal of HHW:

- Read and follow disposal instructions on the product label.

- Never burn or dump any hazardous waste on the ground.

- Do not dispose any hazardous material "down the sink" unless you are sure it can safely be disposed into the sewer system or septic tank.

- Avoid burying any containers or leftover chemicals.

- Do not mix hazardous wastes and do not collect containers and chemicals to dispose of them at one time.

- Solidify any liquid waste. This involves using an absorbent material (dirt, sawdust, rags) to soak up a liquid hazardous material. Do not solidify more than one chemical at a time. Using gloves, sweep or dispose of the material into a plastic bag, and then dispose of with other household garbage.

- Use this same process with any "empty" container other than an aerosol container. It is often good to "open" a non-aerosol container with wire cutters or scissors and air-dry; wearing gloves, swab the inside before disposal. Dispose off the rags or paper towels after they have aired outside.

- With aerosol cans, turn the container upside down and depress spray button, with nozzle facing paper toweling, rags or any absorbent surface. When the spray has lost pressure, wrap the can in several layers of newspaper and dispose with household refuse.

- Latex or water based paint can be solidified by exposing the paint to air. When dried, the paint and container can be disposed with household refuse. Wrap empty containers in several layers of newspaper prior to disposal. This prevents environmental contami-nation and reactive potential.

- Some cleansers can be poured down a drain. If you have a septic tank, drain disposal should nearly always be avoided. If cleansers are designed to be used with water in a home or in sinks, showers, and toilet bowls, the material is probably draining dis-posable. Let the water run, rinse the container and slowly pour the water /chemical down the drain. Allow the water to continue running after the chemical is gone. Allow the container to air dry, wrap in newspaper and dispose in household refuse. • Antifreeze can be flushed down the toilet if connected to a sewer system.

- Pesticides, herbicides, oil paints, paint cleaners, and oil and transmission fluids should never be flushed into a water system or disposed of on ground or put into household refuse. Use these products or share them with someone who can use them.

- Automobile batteries should never be mixed with garbage. Always give those back to battery recyclers.

- Used motor oil should be given to grease manufacturers.

- Batteries should never be thrown in to municipal dumps.

- Note: Make use of any new reuse, recycling, and waste collection and management option for HHP as they become available in your area.

HHP and HHW Management in Industrialized Countries

The United States, Canada and many countries in Europe have banned the sale of certain hazardous products like asbestos, lead-based paint, certain pesticides and mercury containing devices. Many other products like batteries, electronics, carpet and automobiles are required through Extended Producer Responsibility (EPR) laws to be taken back and recycled by the manufacturers. These bans and legislations are instrumental in the protection of human health and the environment.

Household Hazardous Waste is usually collected by municipal solid waste operations in industrialized countries. The two most common ways to collect HHW are through special collection events and permanent collection centres. The special collection events are held for one or two days in a temporary location with hazardous waste transporters on site to take the HHW to a permitted hazardous waste treatment facility at the end of the event.

Permanent collection centres are staffed several days a week to accept and store hazardous waste from homes and small businesses. The waste is packed and shipped to a permitted hazardous waste treatment facilities several times a year.

These centres often have reuse or exchange programs, referred to as "swap shops" in which residents can exchange HHP. For those products covered by EPR laws or that have inherent value (batteries, electronics, used motor oil), retailers act as a collection point.

References

- The-industrial-waste-and-its-effects: earthuntouched.com, Retrieved 15 May 2018

- Industrial-waste-types: byjus.com, Retrieved 25 June 2018

- Characterizing-the-different-types-of-industrial-waste: southernwasteandrecycling.com, Retrieved 15 March 2018

- Hazardous-waste-definition-and-classification-industries-2798: environmentalpollution.in, Retrieved 11 July 2018

Chapter 2

Characteristics of Hazardous Waste

Hazardous wastes are present in the state of solid, liquid or gas. Some of the characteristics of hazardous wastes are reactivity, ignitability, toxicity and corrosivity. This chapter discusses in detail the different properties of hazardous waste for a comprehensive understanding.

Ignitability

ignitable   corrosive   toxic   reactive

A hazardous waste characteristic is a property that indicates that a waste poses a sufficient threat to deserve regulation as hazardous. EPA tried to identify characteristics which, when present in a waste, can cause death or illness in humans or ecological damage. EPA also decided that the presence of any characteristic of hazardous waste should be detectable by using a standardized test method or by applying general knowledge of the waste's properties. EPA believed that unless generators were provided with widely available and uncomplicated test methods for determining whether their wastes exhibited hazardous characteristics, this system of identifying hazardous wastes would be unfair and impractical. Given these criteria, EPA only finalized four hazardous waste characteristics. These characteristics are a necessary supplement to the hazardous waste listings. They provide a screening mechanism that waste handlers must apply to all wastes from all industries. In this sense, the characteristics provide a more complete and inclusive means of identifying hazardous wastes than do the hazardous waste listings. The four characteristics of hazardous waste are:

- Ignitability;

- Corrosivity;

- Reactivity;

- Toxicity.

The regulations explaining these characteristics and the test methods to be used in detecting their presence are found in Part 261, Subpart C. Note that although waste handlers can use the test methods referenced in Subpart C to determine whether a waste displays characteristics, they are not required to do so. In other words, any handler of industrial waste may apply knowledge of the waste's properties to determine if it exhibits a characteristic, instead of sending the waste for expensive laboratory testing. Ignitable, corrosive, and reactive wastes carry the waste codes D001, D002, and D003, respectively. Wastes displaying the characteristic of toxicity can carry any of the waste codes D004 through D043.

Ignitable wastes are wastes that can readily catch fire and sustain combustion. Many paints, cleaners, and other industrial wastes pose such a fire hazard. Most ignitable wastes are liquid in physical form. EPA selected a flash point test as the method for determining whether a liquid waste is combustible enough to deserve regulation as hazardous. The flash point test determines the lowest temperature at which a chemical ignites when exposed to flame. Many wastes in solid or nonliquid physical form (e.g., wood, paper) can also readily catch fire and sustain combustion, but EPA did not intend to regulate most of these nonliquid materials as ignitable wastes. A nonliquid waste is only hazardous due to ignitability if it can spontaneously catch fire under normal handling conditions and can burn so vigorously that it creates a hazard. Certain compressed gases and chemicals called oxidizers can also be ignitable.

Hazardous waste that is classified as ignitable includes the following:

- Liquids with a flashpoint of less than 60° C / 140° F

- Solids that burn spontaneously

- Flammable compressed gas

- Oxidizers

- Materials with National Fire Protection Association (NFPA) or Hazardous Materials Identification System (HMIS) flammability hazard rating of 3 or 4.

Examples of Ignitable Waste

Ignitable liquids: acetone, acetonitrile, benzene, hexane, methanol, ethanol, isopropanol, toluene, xylene, methyl ethyl ketone, lacquer thinner.

Ignitable compressed gases: hydrogen, methane, acetylene, propane, butane, spray-paint cans.

Oxidizers: ammonium persulfate, sodium nitrate, potassium permanganate, sodium perchlorate, hydrogen peroxide (aqueous solution greater than or equal to 8%), potassium peroxide.

Ignitable Wastes

Flashpoint <140° F, or solids that easily catch fire.
- Examples: acetone, xylene, and acetonitrile and other solvents

Compressed flammable gases, such as propane

Flammable solids, such as road flares or carbon.

Ignitable materials are represented by a D001 waste code.

Corrosivity

A substance that is corrosive is one that has the ability to damage or destroy other substances when they come into contact. A corrosive substance can severely attack a number of materials, including organic compounds and metals. In addition to that, living tissues can also be greatly affected by the corrosivity of a substance.

There are different levels of corrosivity. If a corrosive substance is of low concentration, then it is referred to as an irritant. Corrosion of surfaces that are non-living, such as metals, take place in a process that is distinct. A good example is where air or water electrochemical cells eat up or corrode iron, thus resulting in a condition known as rust.

Alkalis, oxidizers, acids, non-alkalis and bases are all known to be corrosive. Poisons are different from corrosives in that the corrosive affects the tissue it comes into contact with immediately. The use of personal protective equipment such as safety shoes, face shields, safety goggles, acid suits, protective gloves and protective aprons, among other personal protective clothing, is recommended when handling chemicals with corrosivity. Most of the protective equipment is made of rubber.

Household cleaning agents are one of the economic advantages of corrosive elements. For instance, in various drain cleaners used for domestic purposes, alkalis or acids are used for the purpose of dissolving proteins and greases that are contained inside the pipes. To increase corrosivity, a catalyst which normally speeds the rate of reaction is used. Once the corrosive has been used, the corrosive elements can either be neutralized or recycled instead of disposing. Untreated corrosives can lead to environmental problems, and thus one should be careful while handling them.

Corrosive wastes are acidic or alkaline (basic) wastes which can readily corrode or dissolve flesh, metal, or other materials. They are also among the most common hazardous wastestreams. Waste sulfuric acid from automotive batteries is an example of a corrosive waste. EPA uses two criteria to identify corrosive hazardous wastes. The first is a pH test. Aqueous wastes with a pH greater than or equal to 12.5, or less than or equal to 2 are corrosive under EPA's rules. A waste may also be corrosive if it has the ability to corrode steel in a specific EPA-approved test protocol. Corrosive wastes carry the waste code D002.

Examples of Corrosive Waste

Corrosive aqueous liquids: hydrochloric acid, sulfuric acid, nitric acid, perchloric acid, acetic acid, sodium hydroxide solution, potassium hydroxide solution.

Reactivity

Reactive wastes are wastes that readily explode or undergo violent reactions. A waste is considered reactive if it:

- Explodes or reacts violently when exposed to water or under normal handling conditions.

- Creates toxic fumes or gases when exposed to water or under common handling conditions Is a cyanide or sulfide bearing waste which, when exposed to pH conditions between 2 and 12.5, can generate toxic gases, vapors or fumes in a quantity sufficient to present a danger to human health or the environment Capable of detonation or explosive reaction if it is subjected to a strong initiating source or if heated under confinement.

- Meets the criteria for classification as a forbidden explosive or a class B explosive under Department of Transportation rules.

The US EPA has developed tests for releasable hydrogen cyanide (HCN) or hydrogen sulfide (H_2S) in SW 846. A waste is reactive if it can generate more than 250 mg HCN or greater than 500 mg H_2S per kilogram of waste.

- The US EPA has rescinded the guidance for HCN and H_2S gas generation, and suggests relying on narrative criteria.

HCN > 250 mg/kg H_2S > 500 mg/kg

According to the HSC, California should use the US EPA thresholds as guidance for reactivity. California does this on a case by case basis and uses narrative criteria to gauge reactivity.

The criteria to identify reactive wastes are purely narrative and descriptive criteria, and can be difficult to test under laboratory conditions. It is fairly straightforward to determine reactivity for pure or relatively pure waste compounds, but mixtures pose a dilemma for two reasons:

- In many cases there are no reliable test methods for mixtures.

- Waste handlers are allowed to use their knowledge about the reactive waste.

These two limitations assume that the dangers posed by reactive wastes are well known to the waste handlers who deal with them.

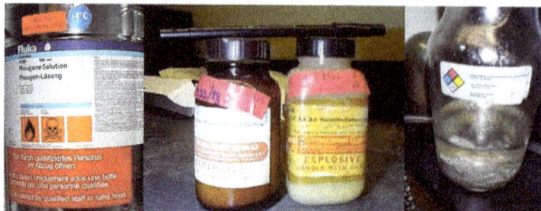

Reactive Wastes   ⦿ VEOLIA
ENVIRONMENTAL SERVICES

Materials that undergo violent change:
- react violently with water or air
- capable of detonation
- create toxic gases when exposed to pH extremes

Examples: sodium metal, extremely dry picric acid, and organic peroxides.

Reactive materials are represented by a D003 waste code.

Toxicity

The leaching of toxic compounds or elements into groundwater drinking supplies from wastes disposed of in landfills is one of the most common ways the general population can be exposed to the chemicals found in industrial wastes. EPA developed a characteristic designed to identify wastes likely to leach dangerous concentrations of certain known toxic chemicals into groundwater. In order to predict whether any particular waste is likely to leach chemicals into groundwater in the absence of special restrictions on its handling, EPA first designed a lab procedure that replicates the leaching process and other effects that occur when wastes are buried in a typical municipal landfill. This lab procedure is known as the Toxicity Characteristic Leaching Procedure (TCLP). Using the TCLP on a waste sample creates a liquid leachate that is similar to the liquid EPA would expect to find in the ground near a landfill containing the same waste. Once the leachate is created in the lab, a waste handler must determine whether it contains any of 39 different toxic chemicals above specified regulatory levels. If the leachate sample contains a sufficient concentration of one of the specified chemicals, the waste exhibits the toxicity characteristic (TC). EPA used groundwater modeling studies and toxicity data for a number of common toxic compounds and elements to set these threshold concentration levels. Much of the toxicity data were originally developed under the Safe Drinking Water Act.

However, there is one exception to using the TCLP to identify a waste as hazardous. The D.C. Circuit Court, in Association of Battery Recyclers vs. EPA, vacated the use of the TCLP to determine whether manufactured gas plant (MGP) wastes exhibit the characteristic of toxicity. As previously stated, the TCLP replicates the leaching process in municipal landfills. The court found that EPA did not produce sufficient evidence that co-disposal of MGP wastes from remediation sites with municipal solid waste (MSW) has happened or is likely to happen. On March 13, 2002, in response to the court vacatur, EPA codified language exempting MGP waste from the toxicity characteristic regulation (67 FR 11251).

To recap, determining whether a waste exhibits the toxicity characteristic involves two principal steps: (1) creating a leachate sample using the TCLP; and (2) evaluating the concentration of 39 chemicals in that sample against the regulatory levels listed below in table. If a waste exhibits the TC, it carries the waste code associated with the compound or element that exceeded the regulatory level. The following table presents the toxicity characteristic waste codes, regulated constituents, and regulatory levels.

Table: Toxicity Characteristic Constituents And Regulatory Levels

Waste Code	Contaminants	Concentration
D004	Arsenic	5.0
D005	Barium	100.0
D018	Benzene	0.5
D006	Cadmium	1.0
D019	Carbon tetrachloride	0.5
D020	Chlordane	0.03
D021	Chlorobenzene	100.0
D022	Chloroform	6.0
D007	Chromium	5.0
D023	o-Cresol*	200.0
D024	m-Cresol*	200.0
D025	p-Cresol*	200.0
D026	Total Cresols*	200.0
D016	2,4-D	10.0
D027	1,4-Dichlorobenzene	7.5
D028	1,2-Dichloroethane	0.5
D029	1,1-Dichloroethylene	0.7
D030	2,4-Dinitrotoluene	0.13
D012	Endrin	0.02
D031	Heptachlor (and its epoxide)	0.008
D032	Hexachlorobenzene	0.13
D033	Hexachlorobutadiene	0.5
D034	Hexachloroethane	3.0
D008	Lead	5.0
D013	Lindane	0.4
D009	Mercury	0.2
D014	Methoxychlor	10.0
D035	Methyl ethyl ketone	200.0
D036	Nitrobenzene	2.0
D037	Pentachlorophenol	100.0
D038	Pyridine	5.0
D010	Selenium	1.0
D011	Silver	5.0

D039	Tetrachloroethylene	0.7
D015	Toxaphene	0.5
D040	Trichloroethylene	0.5
D041	2,4,5-Trichlorophenol	400.0
D042	2,4,6-Trichlorophenol	2.0
D017	2,4,5-TP (Silvex)	1.0
D043	Vinyl chloride	0.2

*If o-, m-, and p-cresols cannot be individually measured, the regulatory level for total Cresols is used.

Chapter 3

Pharmaceutical Waste Management

Biomedical waste is the waste, which contains infectious materials. Any waste of laboratory or medical origin such as microbial cultures, blood, body parts, human or animal tissue, body fluids, etc. can be termed biomedical or pharmaceutical waste. The management of such waste is vital for human health and safety. This chapter closely examines the different types of medical waste, its storage, treatment and disposal.

Pharmaceutical Waste

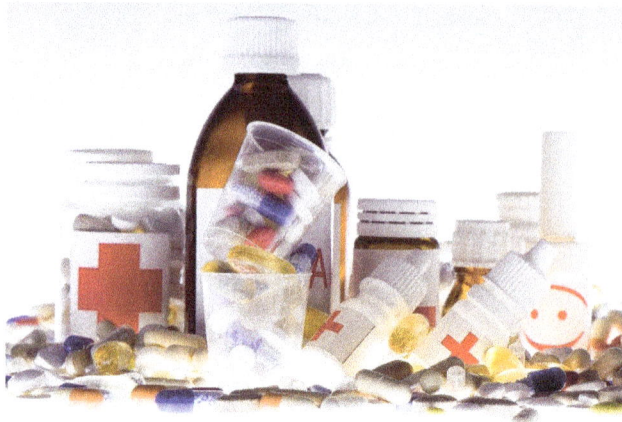

Pharma sector is one of the growing health care sectors that contribute in major way to global economy. Pharmaceuticals that contain active ingredients along with excipients have been found to affect the environment thus public health. The generation of pharmaceutical wastes by pharma companies is enormous irrespective of the size or scale of business. Most common sources of pharmaceutical wastes include disposal by patients and hospitals, livestock feed additives, agricultural wastes due to veterinary use and manufacturing industries. These wastes are reported to present in environment such as drinking waters, surface, waste-water, ground and sewage, and affect human health. Pharmaceuticals or wastes thereof have been observed to reach the environment primarily through usage or inappropriate disposal. Pharmaceutical waste is not one single waste stream, but many distinctive waste streams that can affect the integrity and uniformity of the chemicals that involve pharmaceuticals. This waste is possibly generated through a wide variety of deeds in a healthcare facility, including but not limited to I.V preparation, general compounding, breakages, partially used ampoules, needles, and IVs, out-dated, unused preparations, fallow unit doses, personal medications and out dated pharmaceuticals. Most common sources of pharmaceutical wastes include unhygienic disposal by patients and hospitals, livestock feed additives, agricultural wastes due to veterinary use and manufacturing industries. Their presence in environment such as drinking waters, waste-water, surface, ground and sewage has been reported to affect human as well as aquatic health4 (Heberer,2002). A study reported that the highest levels of organic waste water contaminants detected were from non-prescription drugs, antibiotics and prescription drugs5. Presence of antibiotics such as metronidazole, sulfamethoxazole, ofloxacin, norfloxacin, tinidazole in the range of 1.4 to 236.6 µg-1was documented in hospital effluents in Ujjain District, India6. Drugs like estrogen, ibuprofen, and naproxen have been reported to be in water ways of Los Angeles, even drugs such as aspirin, ibuprofen, indomethacin and carbamazepine have been reported in samples of sewage treatment plants in Canada. In a study, samples taken from the Ohio River contained Escherichia coli with some resistance to penicillin, tetracycline and vancomycin4. (Heberer). World Health Organization documents that around 15% to 35% of hospital waste are infectious and treated as hazardous wastes. Human male sperm counts have dropped 50% on average since 1939 and there have been increases in infertility, cancers caused by hormones, gentital defects and neurological disorders in children . Further the Endocrine disrupters found in US waterways may interfere with normal functioning of the endocrine system, mimic hormones and affect reproduction, development and behaviour. (9.Sheehan E, Wooliever P. Pharmaceutical waste: fish don't need anti-depressants. Pharmacology Associates, LLC). According to a report, about 0.2 to 0.6 kg per capita of waste is generated in Indian cities estimating to around 1.15 lakh metric tonnes of waste per day and 42 million metric tonnes annually8 This emerging issue of pharmaceutical waste throughout India has initiated a need of combined efforts by government entities, health care industries, professionals and even at household level for better management of pharmaceuticals and wastes thereof. There are a number of different options available for the treatment and management of waste containing dodging, minimization, re-use, reutilizing, energy recovery and disposal.

Scientists and religious heads believe that nature was created first and then living beings were introduced to live on nature, which includes the five major components called 'Panchamahabhuta'- water, air, earth, fire and sky. Every living being contains these five elements within its body. Like the need to maintain these elements within body, there is a greater need to protect them in nature too and in the process such maintenance should not become the reason for destruction or pollution.

Waste pharmaceutical agents that are incinerated produce air emissions such as dioxins, acid gases, carbon monoxide, heavy metals and carbon dioxide. These emissions contribute to air pollution, global warming, and are harmful to humans and the environment. Some, such as dioxins, are carcinogens. Moreover, the ash from incineration often must be disposed of as hazardous waste. Despite these concerns, disposal of waste pharmaceuticals via well-managed incineration is the preferred method of disposal at this time. The current best management scenario for waste pharmaceuticals is pollution prevention (P2), a waste management strategy that aims to eliminate and/or reduce the toxicity of the waste10. Pollution prevention in the pharmaceutical industry also entails manufacturers' minimizing waste during production as well as minimizing the level of metabolites excreted from the body. Pharmaceuticals and personal care products can also be procured and distributed in a manner that minimizes waste.

Hazardous Waste is generated on a huge scale by large industries and on a relatively lesser scale by small and medium sized enterprises. Chemicals used for industrial processes often result in formation of dangerous toxic wastes while an admixture of several toxic chemicals and wastes may lead to formation of a lethal synergistic compound. Due to lack of planning, a good portion of the urban population, especially the poorer section often live in the vicinity of illegal Hazardous Waste and municipal solid waste dumpsites, polluted drainage canals and Hazardous Waste generating industries. Risks and threats to public health arising due to improper handling, storage and illegal dumping can be substantially reduced if scientific management practices of waste in designated facilities are adopted. In general, the potentially significant effects of Hazardous Waste on environment includes (i) contamination of soil, ground water and inland surface waters due to percolation of landfill leachate, crude dumping, surface runoff or flood, (ii)bioaccumulation and bio magnification of toxic chemicals in flora and fauna and (iii)gaseous emission from Hazardous Waste incinerators. It is hence imperative that improper handling of harmful materials cause serious problems to the environment and as such, we do need to adopt proper Hazardous Waste management strategies.

The discovery of a variety of pharmaceuticals in surface, ground, and drinking waters around the country is raising concerns about the potentially adverse environmental consequences of these contaminants. Minute concentrations of chemicals known as endocrine disruptors, some of which are pharmaceuticals, are having detrimental effects on aquatic species and possibly on human health and development. The consistent increase in the use of potent pharmaceuticals, driven by both drug development and our aging population, is creating a corresponding increase in the amount of pharmaceutical waste generated. Recent concerns regarding the documentation of drugs in drinking, ground, and surface waters have led to a rapid rise in public awareness and calls for action at the federal, state, and local level. In hospitals, pharmaceutical waste is generally discarded down the drain or land filled, except chemotherapy agents, which are often sent to a regulated medical waste incinerator. These practices were developed at a time when knowledge was not available about the potential adverse effects of introducing waste pharmaceuticals into the environment. Proper pharmaceutical waste management is a highly complex new frontier in environmental management for healthcare facilities. A hospital pharmacy generally stocks between 2,000 and 4,000 different items, each of which must be evaluated against state and federal hazardous waste regulations. Pharmacists and nurses generally do not receive training on hazardous waste management during their academic studies and safety and environmental services managers may not be familiar with the active ingredients and formulations of pharmaceutical products. Frequently used pharmaceuticals, such as physostigmine, warfarin, and chemotherapeutic agents, are regulated as hazardous waste under the Resource Conservation and Recovery Act (RCRA). Failure to comply with hazardous waste regulations by improperly managing and disposing of such waste can result in potentially serious violations and large penalties.

The term Bio Medical Waste means any waste generated in health care processes like diagnosis, treatment or immunization of human beings or animals, research activities concerning production or testing of 'biological'. The Bio-medical Waste (Management and Handling) Rules ("BMW Rules") regulate the manner of disposal of bio-medical wastes ("BM Waste") and provide a detailed framework for the processes and mechanisms to be followed for their effective disposal.

Types of Medical Waste

Biohazardous Waste

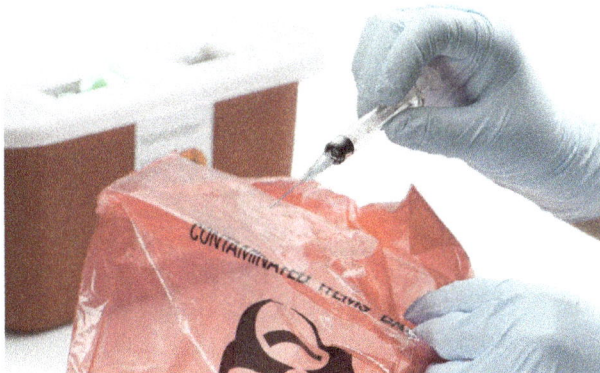

Biohazardous waste is defined as: All biologically contaminated waste that could potentially cause harm to humans, domestic or wild animals or plants. Examples include human and animal blood, tissues, and certain body fluids, recombinant DNA, and human, animal or plant pathogens.

How to handle biohazardous waste: All biohazardous waste must be decontaminated before disposal. Common decontamination methods include heat sterilization (e.g., autoclaving), chemical disinfection, and incineration.

Example of "Waste for Incineration" tag

Animal Carcasses (Including Sheep and Goats; and Any other Animals Infected with Human Pathogens), Tissues, Bedding

1. Collect in leak-proof containers lined with a thick trash bag, label with a "Waste for Incineration" tag, and take to the incinerator at 1676 Veterinary Medicine for incineration.

2. Collect all transgenic animal carcasses or animal carcasses, tissues, and bedding, infected with human pathogens, in leak-proof biohazard bags or containers lined with a thick trash bag.

Liquids

1. Decontaminate all liquid biohazardous materials (such as human blood, bacterial cultures in liquid media, body fluids of animals experimentally infected with pathogens, etc.) by autoclaving or treatment with an appropriate chemical disinfectant for the sufficient contact time.

2. After decontamination, liquids may be disposed of by pouring them down the drain to the sanitary sewer.

    Do Not Put Any Liquids In Regular Trash or Dumpsters

Disposable Solid Items (Non-sharps, and not Animal Carcasses, Tissues or Bedding)

1. Collect all non-sharp disposable items contaminated with biohazardous materials in leak-proof autoclavable biohazard bags (red or orange bag with universal biohazard symbol or other bag tagged with red or orange universal biohazard symbol). Before decontaminating, place an autoclave indicator tape "X" over the biohazard symbol. Decontaminate the bags by autoclaving for a minimum of 45 minutes before disposal.

Examples

Biohazard bags

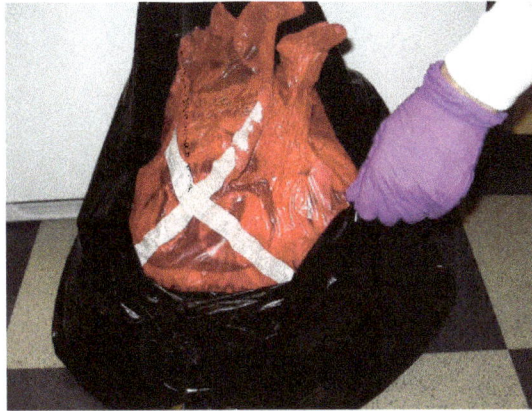

Biohazard symbol with "X"

After autoclaving, place the now decontaminated biohazard bag into a dark garbage bag, seal, and place in regular trash.

Non-disposable or Reusable Items

Decontaminate non-disposable or reusable items contaminated with biohazardous materials by using a chemical disinfectant. Choose a chemical disinfectant appropriate for the specific biohazardous material being used and allow for sufficient contact time.

Metal Sharps

1. Use separate containers for metal, glass, and plastic sharps. Collect all metal sharps contaminated with biohazardous materials in leak-proof, puncture resistant containers which have been labeled with the universal biohazard symbol. Decontaminate the containers by autoclaving.

NOTE: To prevent needle sticks, do NOT recap the needles or remove from syringes, instead discard the entire unit into the sharps waste container designated for biohazardous sharps.

2. After autoclaving, label the now decontaminated sharps waste containers with a "Non-Infectious Syringes and Metal Sharps Only" label.

3. Collect metal sharps that have never been contaminated with biohazardous materials (e.g., used only with chemicals) in leak-proof, puncture resistant white plastic containers labeled with a "Non-infectious Syringes and Metal Sharps Only" label. Do not autoclave these containers, because they will melt.

Label Examples

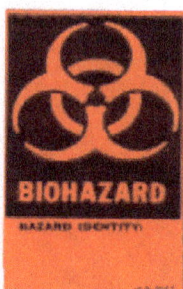

Glass Sharps

1. Use separate containers for metal, glass, and plastic sharps. Collect all glass sharps contaminated with biohazardous materials in leak proof, puncture resistant containers which have been labeled with the universal biohazard symbol. Decontaminate the containers by autoclaving.

2. After autoclaving, empty the now decontaminated glass sharps container into a yellow tidy cat container in your laboratory for storage or into the yellow glass disposal bin on your building's loading dock for disposal.

3. Collect glass sharps that have never been contaminated with biohazardous materials in a yellow tidy cat container in your laboratory for storage or into the yellow glass disposal bin on your building's loading dock for disposal. Autoclaving of these containers is not necessary.

Plastic Sharps

1. Use separate containers for metal, glass, and plastic sharps. Collect plastic materials (pipette tips, plastic pipettes) that can poke out of bags and contaminated with biohazardous materials in leak-proof, puncture resistant containers

which have been labeled with the universal biohazard symbol. Decontaminate the containers by autoclaving.

2. After autoclaving, place the now decontaminated plastic sharps inside a garbage bag lined cardboard box, seal, label "Plastic Sharps" and throw into the regular trash dumpster.

Infectious Medical Waste

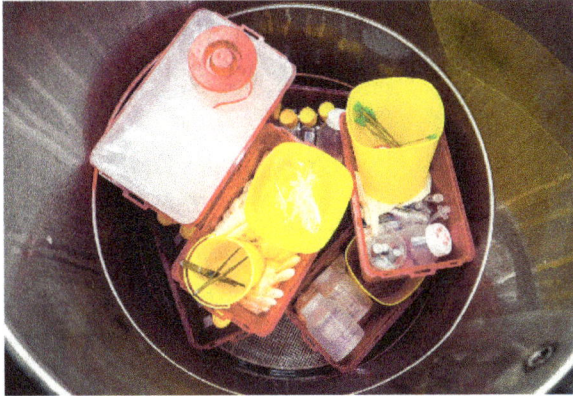

Infectious waste is municipal and residual waste which is generated in the diagnosis, treatment, immunization or autopsy of human beings or animals, in research pertaining thereto, in the preparation of human or animal remains for interment or cremation, or in the production or testing of biological, and which falls under one or more of the following categories:

- Cultures and stocks of infectious agents and associated biologicals, including the following: cultures from medical and pathological laboratories; cultures and stocks of infectious agents from research and industrial laboratories; wastes from the production of biologicals; discarded live and attenuated vaccines except for residue in emptied containers; and culture dishes, assemblies and devices used to conduct diagnostic tests or to transfer, inoculate and mix cultures.

- Pathological wastes: Human pathological wastes, including tissues, organs and body parts and body fluids that are removed during surgery, autopsy, other medical procedures or laboratory procedures. The term does not include hair, nails or extracted teeth.

- Human blood and body fluid waste:

    ◦ Liquid waste human blood.

    ◦ Blood products.

    ◦ Items saturated or dripping with human blood.

- Items that were saturated or dripping with human blood that are now caked with dried human blood, including serum, plasma and other blood components, which were used or intended for use in patient care, specimen testing or the development of pharmaceuticals.

- Intravenous bags that have been used for blood transfusions.

- Items, including dialysate, that have been in contact with the blood of patients undergoing hemodialysis at hospitals or independent treatment centres.

- Items saturated or dripping with body fluids or caked with dried body fluids from persons during surgery, autopsy, other medical procedures or laboratory procedures.

- Specimens of blood products or body fluids, and their containers.

- Animal wastes: Contaminated animal carcasses, body parts, blood, blood products, secretions, excretions and bedding of animals that were known to have been exposed to zoonotic infectious agents or nonzoonotic human pathogens during research (including research in veterinary schools and hospitals), production of biologicals or testing of pharmaceuticals.

- Isolation wastes: Biological wastes and waste contaminated with blood, excretion, exudates or secretions from:

  - Humans who are isolated to protect others from highly virulent diseases.

  - Isolated animals known or suspected to be infected with highly virulent diseases.

- Used sharps: Sharps that have been in contact with infectious agents or that have been used in animal or human patient care or treatment, at medical, research or industrial laboratories.

Exceptions to the definition of infectious waste include:

- Wastes generated as a result of home self-care.

- Human corpses, remains and anatomical parts that are intended for interment or cremation, or are donated and used for scientific or medical education, research or treatment.

- Etiologic agents being transported for purposes other than waste processing or disposal pursuant to the requirements of the United States Department of Transportation, the Pa. Department of Transportation and other applicable shipping requirements.

- Samples of infectious waste transported offsite by Commonwealth or United States government enforcement personnel during an enforcement proceeding.

- Body fluids or biologicals which are being transported to or stored at a laboratory prior to laboratory testing.

- Ash residue from the incineration of materials if the incineration was conducted in accordance with infectious waste monitoring requirements. The ash residue shall be managed as special handling municipal waste.

- Reusable or recyclable containers or other nondisposable materials, if they are cleaned and disinfected, or if there has been no direct contact between the surface of the container and materials. Laundry or medical equipment shall be cleaned and disinfected in accordance with the U.S. Occupational Safety and Health Administration Requirements relating to blood borne pathogens.

- Soiled diapers.

- Mixtures of hazardous waste and other materials identified in the regulations shall be managed as hazardous waste and not infectious waste.

- Mixtures of materials identified in the regulations and regulated radioactive waste shall be managed as radioactive waste in accordance with applicable Commonwealth and Federal statutes and regulations.

- Mixtures of materials identified in the regulations and chemotherapeutic waste shall be managed as chemotherapeutic waste.

Managing Infectious Waste

Basic Storage Requirements

Infectious and chemotherapeutic waste shall be stored and contained in a manner that:

- Maintains the integrity of the containers, prevents the leakage or release of waste from the containers and provides protection from water, rain and wind;

- Prevents the spread of infectious or chemotherapeutic agents;

- Affords protection from animals and does not provide a breeding place or a food source for insects or rodents;

- Maintains the waste in a nonputrescent state;

- Prevents odors from emanating from the container; and

- Prevents unauthorized access to the waste. As part of this requirement, the following shall be met:

  ○ Enclosures and containers used for storage of infectious or chemotherapeutic waste shall be secured to deny access to unauthorized persons. Enclosures and containers shall also be marked with prominent warning signs indicating the storage of infectious or chemotherapeutic waste.

- Enclosures at a waste generating or processing facility that are used for the storage of infectious or chemotherapeutic waste shall be constructed of finish materials that are impermeable and capable of being readily maintained in a sanitary condition. Storage areas shall be ventilated to minimize human exposure to the exhaust air.

- Infectious and chemotherapeutic waste may not be commingled with other waste.

- The generator may store infectious and municipal waste that has been sorted and separately containerized on the same cart for movement to an onsite processing or disposal facility. Chemotherapeutic waste may also be stored on the cart with municipal and infectious waste if it is sorted.

Sorting

Infectious and chemotherapeutic waste shall be placed in separate containers from other waste at the point of origin in the generating facility. Infectious and chemotherapeutic waste may be stored together in the same container if approved in writing by the Department.

Used sharps, regardless of whether they are infectious or chemotherapeutic waste, may be stored in the same container

Infectious waste shall be sorted at the point of origin in the generating facility into the following three classes, and each class shall be placed in a separate container:

- Used sharps;
- Fluid quantities greater than 20 cubic centimetres; and
- Other infectious waste.

Chemotherapeutic waste shall also be sorted at the point of origin in the generating facility into the same classes. However, fluids must be separated regardless of the volume.

Sorted and separately containerized infectious waste may be placed together into another container for onsite handling or offsite transportation.

Storage for Infectious Waste for Generators

Generators of infectious or chemotherapeutic waste may store the waste onsite according to the following requirements:

- Infectious waste, excluding used sharps, may be stored at room temperature until the storage container is full, but for no longer than 30 days from the date waste was first placed in the container. If the infectious waste becomes putrescent during the storage period, the waste shall be moved offsite within 24 hours for processing or disposal.

- A storage container filled with infectious waste may be stored in a refrigeration unit for up to 30 days from the date waste was first placed in the container.

- A storage container of infectious waste that has been filled within 30 days from the date waste was first placed in the container may be frozen immediately for up to 90 days from the date waste was first placed in the container.

Used sharps containers may be used until full as long as the storage is in accordance with basic storage requirements.

Storage for Infectious Waste Processors

If the waste processing facility is separate from the waste generating facility, infectious waste may not be stored at the waste processing facility for more than the following periods unless other periods are approved in a permit:

- Seventy-two hours at a temperature <=28°C;

- Seven days in a refrigerator at <=7°C; or

- Thirty days in a freezer at <=-18°C.

Storage Containers

Infectious and chemotherapeutic waste shall be placed in containers that are leak proof; impervious to moisture; and sufficient in strength to prevent puncturing, tearing or bursting during storage.

- In addition to the above requirements, used sharps shall be stored in containers that are rigid; tightly lidded; and puncture resistant.

- In addition, infectious waste fluids in quantities greater than 20 cubic centimetres, and chemotherapeutic waste fluids shall be stored in containers that are break-resistant; and tightly lidded or tightly stoppered.

When bags are used as the only storage container, double or multiple bagging shall be used and the following requirements shall be met:

Upon packaging, bags shall be securely tied.

The bag shall be constructed of material of sufficient single thickness strength to meet DEP requirements and ASTM Standards. The bags shall be certified.

Marking of Containers

The outermost container for each package of infectious or chemotherapeutic waste for offsite transportation shall be labeled immediately after packing. The label shall be securely attached and be clearly legible. Indelible ink shall be used to complete the information on the label. If handwritten, the label shall be at least 3 inches by 5 inches in dimension. The following information shall be included on the label:

- Name, address and telephone number of the generator;

- Date the waste was generated; and

- The name of the transporter and, if applicable, Department-issued infectious and chemotherapeutic waste transporter license number.

The following information shall be printed on the outermost container or bag for each package of infectious or chemotherapeutic waste for either onsite movement or offsite transportation:

- The words "infectious waste" or "chemotherapeutic waste," whichever is applicable; and

- The universal biohazard symbol.

A container used for infectious waste cannot be used again unless it has been decontaminated, or the container surface has been protection from direct contact with the waste.

Transportation

Only DEP-licensed transporters may transport infectious and chemotherapeutic waste. Transporters must follow strict requirements in regard to containment and packaging, in order to maintain the integrity of the containers and to prevent leakages, spills and releases of the waste. Transporters must also comply with annual reporting requirements.

Permit-by Rule—On-Site Processing Facilities

A majority of infectious and chemotherapeutic waste on-site processing facilities may receive a permit-by-rule, if they meet the following conditions:

- The facility complies with all applicable municipal waste storage, collection and transportation requirements;

- The facility has all required and necessary permits;

- The facility operator maintains all required plans, records, disposal and other DEP-required information in a readily accessible location;

- Waste processing does not adversely affect the public health, safety, welfare or the environment;

- Waste is properly disinfected; and

- A log with required information is kept for each disinfection unit.

Potentially Infectious Medical Waste

PIMW is referred to by many different terms including medical, infectious, red-bag, hospital, biohazardous, and regulated waste. This waste can be defined as any waste containing sufficient virulence and quantity so that exposure to the waste by susceptible individuals could result in an infectious disease.

Waste Comes From

PIMW is generated in the clinical and research laboratories as well as in the Hospital and Clinics. Waste sources may come from:

- The diagnosis, treatment or immunization of human beings or animals.

- Biological or medical research and education laboratories.

- The production or testing of biologicals.

PIMW includes articles such as contaminated gauze, disposable gloves and ALL needles and syringes etc.

The Guidelines for the Safe Handling and Disposal of PIMW

- Segregation - It is the responsibility of the generator to segregate PIMW from ordinary waste. The segregation should occur at the point of generation (where materials become waste). The generator of PIMW must segregate PIMW into solid and liquid (such as blood and body fluids) waste. Bulk blood, suctioned fluids, excretions, and secretions may be carefully poured down a drain connected to a sanitary sewer and flushed thoroughly. To adequately segregate waste, generators of PIMW must provide at least two receptacles for waste: one for ordinary waste and one for PIMW. The receptacles intended for PIMW must be red and labeled with the Universal biohazard symbol. Never use a red-labeled bag unless it is inside a container.

- Containment – The disposal of PIMW must not compromise the strength and durability of the red-labeled plastic bag. It is imperative that objects that could potentially tear or puncture the bag be packaged in such a manner to prevent such an occurrence. It is extremely important that they not be overfilled. A properly filled bag will allow the opening to be easily pulled closed and knotted, sealed, or twist-tied. If additional strength is necessary or if the bag becomes contaminated, PIMW should be double bagged.

Handling Potentially Infectious Medical Waste

These Items DO Go Into the Red Bag	These Items DON'T Go Into the Red Bag
• Blood & Body Fluids	• Compressed Gas Cylinders
• Blood Saturated Items	• Fixatives & Preservatives
• Saturated Bandages	• Garbage
• Saturated Gauze	• Hazardous & Chemical Waste
• Visibly Bloody Gloves	• Loose Sharps
• Visibly Bloody Plastic Tubing	• Medications (Chemotherapeutic Agents)
• Visibly Contaminated PPE	• Radioactive Waste

- Handling and Transport – The disposal of PIMW shall be in accordance with the department established to service the location of the generator (Building Services or Housekeeping). When removing bags from receptacles: immediately knot, seal, or twist-tie the bag closed. Never transport a bag that is not closed. When handling, avoid contact with skin, clothing, furniture, building fixtures, walls and floors. It is recommended that Personal Protective Equipment (PPE), such as rubber gloves and an apron be worn. ALWAYS wash hands immediately after handling. The use of carts or containers is recommended to prevent damage or spillage.

- Spills and clean up – In the event of a waste spill, use PPE and carefully clean and disinfect the affected area. Proper techniques and an approved disinfectant (e.g., a 1:10 bleach solution) should be used. Alternately, the housekeeping or building service personnel assigned to the area should be contacted. Avoid all foot traffic through the spill and confine the area. When there is a delay in spill clean-up, the responsible party should remain at the spill to provide assistance.

About Disposal of Sharps

A "sharp" is any object that is capable of puncturing the skin. Sharps include needles, syringes, contaminated broken glass, scalpels, culture slides, contaminated glass culture dishes, capillary tubes/pipettes, and micropipette tips, contaminated broken rigid plastic and exposed ends of dental wires. All sharps contaminated with infectious materials and ALL needles and syringes will be disposed of following these guidelines.

- Never bend, clip, deform, or break a needle in any manner. Devices for such purposes shall not be used.

- Never recap or resheath a needle after the protective covering has been removed.

- Deposit sharps directly into rigid, puncture- and leak-resistant containers that are designated for sharps waste. The container should be labeled with the Universal Biohazard symbol.

- Always maintain sharp containers in an upright position.

- Never overfill sharps containers. Once filled, containers must have all openings securely closed.

- If leakage from a sharp container is possible, the container must be placed inside another leak-proof labeled container.

- Never abandon sharps and allow them to remain unattended on furniture surfaces such as countertops, lab benches, chairs etc.

- Never place sharps in the regular trash or in biohazard bags.

For Supplies Required For Biohazardous Waste Handling and Disposal

Supplies Provider	Contact
Campus-wide	Fisher Scientific ext. 3-0324
Hospital and Clinics	Material Management ext. 6-3682

For Biohazardous Waste Pickup

Location	Contact
Hospital and Clinics	Hospital Environmental Services ext. 6-3688
West Side	Building Services ext. 6-7468
East Side	Building Services ext. 6-1799
Out-Patient Clinics (Buildings 902, 950 and 957)	Building Services ext. 6-7468
College of Dentistry (#940)	Fred Chappa ext. 6-7633
MBRB (#919)	Bernie Greski ext. 6-6963 (problems only)
BRL (#932)	Scott Hauff ext. 6-7052
School of Public Health, West (#930)	Margit Javor ext. 3-1390
College of Medicine Research Building (#934)	Building Services ext. 6-7468 or Anthony Capistran 3-1551

Healthcare Waste

Process of Disposal and Treatment of Health Care Waste

Health-care waste includes all the waste generated by health-care establishments, re-search facilities, and laboratories. In addition, it includes the waste originating from "minor" or "scattered" sources—such as that produced in the course of health care undertaken in the home.

Between 75% and 90% of the waste produced by health-care providers is non-risk or "general" health-care waste, comparable to domestic waste. It comes mostly from the administrative and housekeeping functions of health-care establishments and may also include waste generated during maintenance of health-care premises. The remaining 10–25% of healthcare waste is regarded as hazardous and may create a variety of health risks. This handbook is concerned almost exclusively with hazardous health-care waste general wastes should be dealt with by the municipal waste disposal mechanisms.

Infectious Waste

Infectious waste is suspected to contain pathogens (bacteria, viruses, parasites, or fungi) in sufficient concentration or quantity to cause disease in susceptible hosts. This category includes:

- cultures and stocks of infectious agents from laboratory work;

- waste from surgery and autopsies on patients with infectious diseases (e.g. tissues, and materials or equipment that have been in contact with blood or other body fluids);

- waste from infected patients in isolation wards (e.g. excreta, dressings from infected or surgical wounds, clothes heavily soiled with human blood or other body fluids);

- waste that has been in contact with infected patients undergoing haemodialysis (e.g. dialysis equipment such as tubing and filters, disposable towels, gowns, aprons, gloves, and laboratory coats);

- infected animals from laboratories;

- any other instruments or materials that have been in contact with infected persons or animals.

Categories of Health-care Waste

Waste category	Description and examples
Infectious waste	Waste suspected to contain pathogens e.g. laboratory cultures; waste from isolation wards; tissues (swabs), materials, or equipment that have been in contact with infected patients; excreta
Pathological waste	Human tissues or fluids e.g. body parts; blood and other body fluids; fetuses
Sharps	Sharp waste e.g. needles; infusion sets; scalpels; knives; blades; broken glass
Pharmaceutical waste	Waste containing pharmaceuticals e.g. pharmaceuticals that are expired or no longer needed; items contaminated by or containing pharmaceuticals (bottles, boxes)
Genotoxic waste	Waste containing substances with genotoxic properties e.g. waste containing cytostatic drugs (often used in cancer therapy); genotoxic chemicals
Chemical waste	Waste containing chemical substances e.g. laboratory reagents; film developer; disinfectants that are expired or no longer needed; solvents
Wastes with high content of heavy metals	Batteries; broken thermometers; blood-pressure gauges; etc.
Pressurized containers	Gas cylinders; gas cartridges; aerosol cans
Radioactive waste	Waste containing radioactive substances e.g. unused liquids from radiotherapy or laboratory research; contaminated glassware, packages, or absorbent paper; urine and excreta from patients treated or tested with unsealed radionuclides; sealed sources

Cultures and stocks of highly infectious agents waste from autopsies, animal bodies, and other waste items that have been inoculated, infected, or in contact with such agents are called highly infectious waste.

Pathological Waste

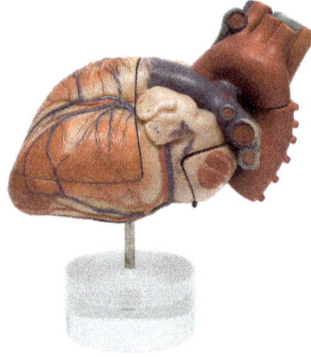

Pathological waste consists of tissues, organs, body parts, human foetuses and animal carcasses, blood, and body fluids. Within this category, recognizable human or animal body parts are also called anatomical waste. This category should be considered as a subcategory of infectious waste, even though it may also include healthy body parts.

Sharps

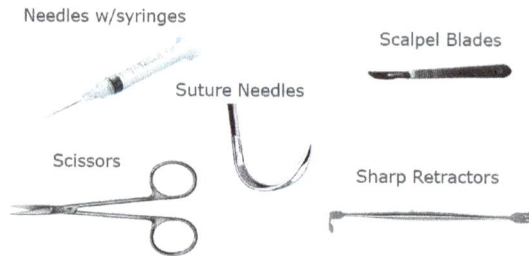

Sharps are items that could cause cuts or puncture wounds, including needles, hypodermic needles, scalpel and other blades, knives, infusion sets, saws, broken glass, and nails. Whether or not they are infected, such items are usually considered as highly hazardous health-care waste.

Pharmaceutical Waste

Pharmaceutical waste includes expired, unused, spilt, and contaminated pharmaceutical products, drugs, vaccines, and sera that are no longer required and need to be disposed of appropriately. The category also includes discarded items used in the handling of pharmaceuticals, such as bottles or boxes with residues, gloves, masks, connecting tubing, and drug vials.

Genotoxic Waste

Cytotoxic & Genotoxic Waste

All chemotherapy medicine items, specimen tubes.

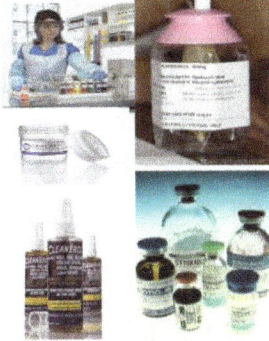

Dispose in Yellow bag

Genotoxic waste is highly hazardous and may have mutagenic, teratogenic, or carcinogenic properties. It raises serious safety problems, both inside hospitals and after disposal, and should be given special attention. Genotoxic waste may include certain cytostatic drugs, vomit, urine, or faeces from patients treated with cytostatic drugs, chemicals, and radioactive material.

Cytotoxic (or antineoplastic) drugs, the principal substances in this category, have the ability to kill or stop the growth of certain living cells and are used in chemotherapy of cancer. They play an important role in the therapy of various neoplastic conditions but are also finding wider application as immunosuppressive agents in organ transplantation and in treating various diseases with an immunological basis. Cytotoxic drugs are most often used in specialized departments such as oncology and radiotherapy units, whose main role is cancer treatment; however, their use in other hospital departments is increasing and they may also be used outside the hospital setting.

The most common genotoxic substances used in health care are listed.

Most Common Genotoxic Products used in Health Care

Classified as Carcinogenic

Chemicals:

benzene

Cytotoxic and other drugs:

azathioprine, chlorambucil, chlornaphazine, ciclosporin, cyclophosphamide, melphalan, semustine, tamoxifen, thiotepa, treosulfan

Radioactive substances:

>   (radioactive substances are treated as a separate category in this handbook)

Classified as Possibly or Probably Carcinogenic

Cytotoxic and other drugs:

>   azacitidine, bleomycin, carmustine, chloramphenicol, chlorozotocin, cisplatin, dacarbazine, daunorubicin, dihydroxymethylfuratrizine (e.g. Panfuran S—no longer in use), doxorubicin, lomustine, methylthiouracil, metronidazole, mitomycin, nafenopin, niridazole, oxazepam, phenacetin, phenobarbital, phenytoin, procarbazine hydrochloride, progesterone, sarcolysin, streptozocin, trichlormethine.

Harmful cytostatic drugs can be categorized as follows:

- alkylating agents: cause alkylation of DNA nucleotides, which leads to cross-linking and miscoding of the genetic stock;

- antimetabolites: inhibit the biosynthesis of nucleic acids in the cell;

- mitotic inhibitors: prevent cell replication.

Cytotoxic wastes are generated from several sources and can include the following:

- contaminated materials from drug preparation and administration, such as syringes, needles, gauges, vials, packaging;

- outdated drugs, excess (leftover) solutions, drugs returned from the wards;

- urine, faeces, and vomit from patients, which may contain potentially hazardous amounts of the administered cytostatic drugs or of their metabolites and which should be considered genotoxic for at least 48 hours and sometimes up to 1 week after drug administration.

In specialized oncological hospitals, genotoxic waste (containing cytostatic or radioactive substances) may constitute as much as 1% of the total health-care wastes.

Chemical Waste

Chemical waste consists of discarded solid, liquid, and gaseous chemicals, for example from diagnostic and experimental work and from cleaning, housekeeping, and disinfecting procedures. Chemical waste from health care may be hazardous or non-hazardous; in the context of protecting health, it is considered to be hazardous if it has at least one of the following properties:

- toxic;

- corrosive (e.g. acids of pH < 2 and bases of pH > 12);

- flammable;

- reactive (explosive, water-reactive, shock-sensitive);

- genotoxic (e.g. cytostatic drugs).

Non-hazardous chemical waste consists of chemicals with none of the above properties, such as sugars, amino acids, and certain organic and inorganic salts.

The types of hazardous chemicals used most commonly in maintenance of health-care centres and hospitals and the most likely to be found in waste are discussed in the following paragraphs.

Formaldehyde

Formaldehyde is a significant source of chemical waste in hospitals. It is used to clean and disinfect equipment (e.g. haemodialysis or surgical equipment), to preserve specimens, to disinfect liquid infectious waste, and in pathology, autopsy, dialysis, embalming, and nursing units.

Photographic Chemicals

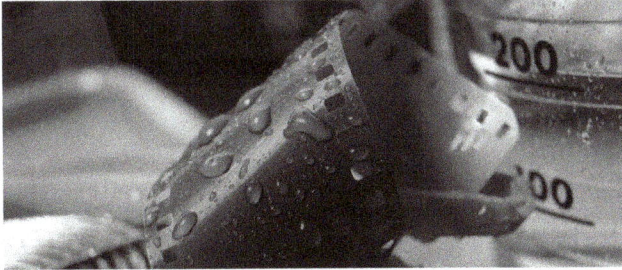

Photographic fixing and developing solutions are used in X-ray departments. The fixer usually contains 5–10% hydroquinone, 1–5% potassium hydroxide, and less than 1% silver. The developer contains approximately 45% glutaraldehyde. Acetic acid is used in both stop baths and fixer solutions.

Solvents

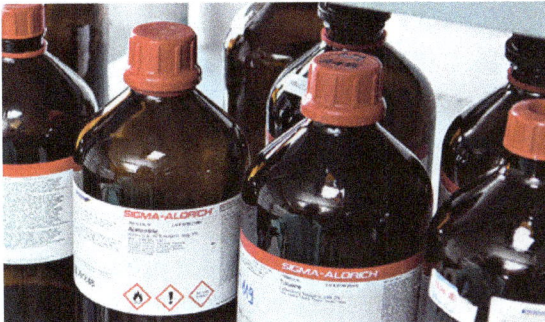

Wastes containing solvents are generated in various departments of a hospital, including pathology and histology laboratories and engineering departments. Solvents used in hospitals include halogenated compounds, such as methylene chloride, chloroform, trichloroethylene, and refrigerants, and non-halogenated compounds such as xylene, methanol, acetone, isopropanol, toluene, ethyl acetate, and acetonitrile.

Organic Chemicals

Waste organic chemicals generated in health-care facilities include:

- disinfecting and cleaning solutions such as phenol-based chemicals used for scrubbing floors, perchlorethylene used in workshops and laundries;

- oils such as vacuum-pump oils, used engine oil from vehicles (particularly if there is a vehicle service station on the hospital premises);

- insecticides, rodenticides.

Inorganic Chemicals

Waste inorganic chemicals consist mainly of acids and alkalis (e.g. sulfuric, hydrochloric, nitric, and chromic acids, sodium hydroxide and ammonia solutions). They also include oxidants, such as potassium permanganate ($KMnO_4$) and potassium dichromate ($K_2Cr_2O_7$), and reducing agents, such as sodium bisulfite ($NaHSO_3$) and sodium sulphite (Na_2SO_3).

Wastes with high Content of Heavy Metals

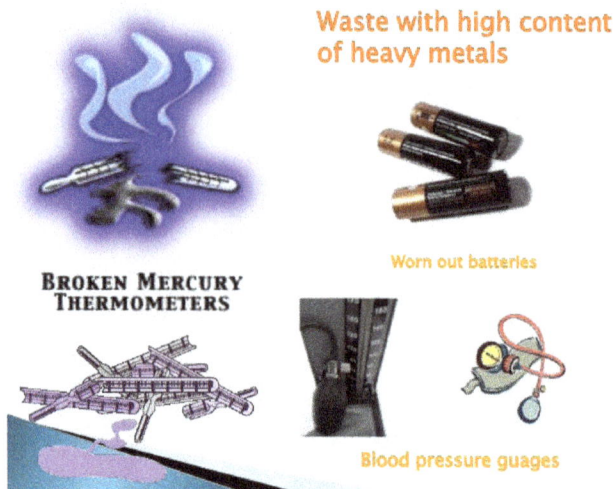

Waste with high content of heavy metals

BROKEN MERCURY THERMOMETERS

Worn out batteries

Blood pressure guages

Wastes with a high heavy-metal content represent a subcategory of hazardous chemical waste, and are usually highly toxic. Mercury wastes are typically generated by spillage from broken clinical equipment but their volume is decreasing with the substitution of solid-state electronic sensing instruments (thermometers, blood-pressure gauges, etc.). Whenever possible, spilled drops of mercury should be recovered. Residues from dentistry have a high mercury content. Cadmium waste comes mainly from discarded batteries. Certain "reinforced wood panels" containing lead is still used in radiation proofing of X-ray and diagnostic departments. A number of drugs contain arsenic, but these are treated here as pharmaceutical waste.

Pressurized Containers

Many types of gas are used in health care, and are often stored in pressurized cylinders, cartridges, and aerosol cans. Many of these, once empty or of no further use (although they may still contain residues), are reusable, but certain types—notably aerosol cans— must be disposed of.

Whether inert or potentially harmful, gases in pressurized containers should always be handled with care; containers may explode if incinerated or accidentally punctured.

Most Common Gases used in Health Care

Anaesthetic Gases

nitrous oxide, volatile halogenated hydrocarbons (such as halothane, isoflurane, and enflurane), which have largely replaced ether and chloroform.

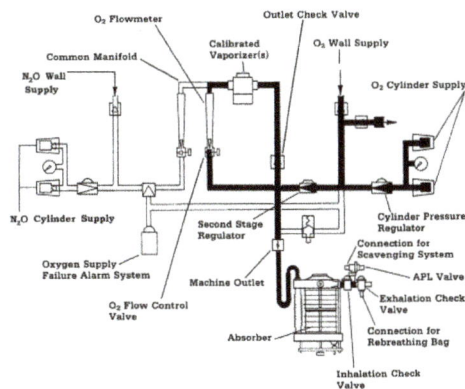

Applications—in hospital operating theatres, during childbirth in maternity hospitals, in ambulances, in general hospital wards during painful procedures, in dentistry, for sedation, etc.

Ethylene Oxide

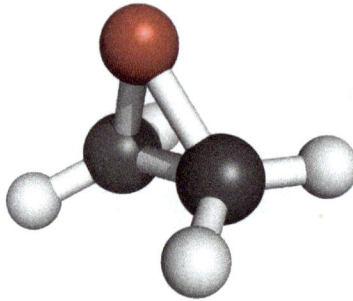

Applications—for sterilization of surgical equipment and medical devices, in central supply areas, and, at times, in operating rooms.

Oxygen

Stored in bulk tank or cylinders, in gaseous or liquid form, or supplied by central piping.

Application—inhalation supply for patients.

Compressed air

Applications—in laboratory work, inhalation therapy equipment, maintenance equipment, and environmental control systems.

Radioactive Waste

Background on Radioactivity

Ionizing radiations cannot be detected by any of the senses and—other than burns, which may occur in exposed areas—usually cause no immediate effects unless an individual receives a very high dose. The ionizing radiations of interest in medicine include the X-rays, a- and b-particles, and g-rays emitted by radioactive substances. An important practical difference between these types of radiation is that X-rays from X-ray tubes are emitted only when generating equipment is switched on, whereas radiation from radionuclides can never be switched off and can be avoided only by shielding the material.

Radionuclides continuously undergo spontaneous disintegration (known as "radioactive decay") in which energy is liberated, generally resulting in the formation of new nuclides. The process is accompanied by the emission of one or more types of radiation, such as a- and b-particles and g-rays. These cause ionization of intracellular material; radioactive substances are therefore genotoxic.

- α-Particles are heavy, positively charged, and include protons and neutrons. They have a low penetration power, and are hazardous to humans mostly when inhaled or ingested.

- β-Particles are negatively or positively charged electrons with significant ability to penetrate human skin; they affect health through ionization of intracellular proteins and proteinaceous components.

- γ-Rays are electromagnetic radiations similar to X-rays but of shorter wavelength. Their penetrating power is high and lead (or thick concrete) shielding is required to reduce their intensity.

Disintegration is measured in terms of the time required for the radioactivity to decrease by half—the "half-life". Each radionuclide has a characteristic half-life, which is constant and by which it may be identified. Half-lives range from fractions of a second to millions of years. Values for the most common radionuclides used in nuclear medicine are listed in Table.

The activity of a radioactive substance corresponds to the disintegration rate and is measured in becquerels (Bq), the SI unit that has replaced the curie (Ci):

1Bq = 1 disintegration per second

$1Ci = 3.7 \times 10^{10}Bq$

The amount of energy absorbed, per unit mass, as a result of exposure to ionizing radiation is called the absorbed dose and is expressed in gray (Gy); this SI unit has replaced the rad (1Gy = 100rad). However, different types of radiation have different effects according to the biological material and the type of tissue. To allow for these differences, absorbed dose is averaged over an organ or tissue and "weighted" for the type of radiation. This yields the equivalent dose, measured in sievert (Sv), which replaces the rem (1Sv = 100rem).

Radioactive Substances used in Health Care and Generating Waste

Radioactive waste includes solid, liquid, and gaseous materials contaminated with radionuclides. It is produced as a result of procedures such as in-vitro analysis of body tissue and fluid, in-vivo organ imaging and tumour localization, and various investigative and therapeutic practices.

Principal radionuclides used in health-care establishments:

Radionuclide[b]	Emission	Format	Half-life	Application
^{3}H	β	Unsealed	12.3 years	Research
^{14}C	β	Unsealed	5730 years	Research
^{32}P	β	Unsealed	14.3 days	Diagnosis; therapy
^{51}Cr	γ	Unsealed	27.8 days	*In-vitro* diagnosis
^{57}Co	β	Unsealed	271 days	*In-vitro* diagnosis
^{60}Co	β	Sealed	5.3 years	Diagnosis; therapy; research
^{59}Fe	β	Unsealed	45 days	*In-vitro* diagnosis
^{67}Ga	γ	Unsealed	78 hours	Diagnostic imaging
^{75}Se	γ	Unsealed	119 days	Diagnostic imaging
^{85}Kr	β	Unsealed	10.7 years	Diagnostic imaging; research
^{99m}Tc	γ	Unsealed	6 hours	Diagnostic imaging
^{123}I	γ	Unsealed	13.1 hours	Diagnostic uptake; therapy
^{125}I	γ	Unsealed	60 days	Diagnostic uptake; therapy
^{131}I	β	Unsealed	8 days	Therapy
^{133}Xe	β	Unsealed	5.3 days	Diagnostic imaging
^{137}Cs	β	Sealed	30 years	Therapy; research
^{192}Ir	β	Sealed (ribbons)	74 days	Therapy
^{198}Au	β	Sealed (seeds)	2.3 days	Therapy
^{222}Rd	α	Sealed (seeds)	3.8 days	Therapy
^{226}Ra	α	Sealed	1600 years	Therapy

[a]Adapted from WHO (1985).

[b]3H and 14C used for research purposes account for the largest amount of radioactive health-care waste.

Radionuclides used in health care are usually conditioned in unsealed (or "open") sources or sealed sources. Unsealed sources are usually liquids that are applied directly and not encapsulated during use; sealed sources are radioactive substances contained in parts of equipment or apparatus or encapsulated in unbreakable or impervious objects such as "seeds" or needles.

Radioactive health-care waste usually contains radionuclides with short half-lives, which lose their activity relatively quickly. Certain therapeutic procedures, however, require the use of radionuclides with longer half-lives; these are usually in the form of pins, needles, or "seeds" and may be reused on other patients after sterilization.

The type and form of radioactive material used in health-care establishments usually results in low-level radioactive waste (<1MBq). Waste in the form of sealed sources may be of fairly high activity, but is only generated in low volumes from larger medical and research laboratories. Sealed sources are generally returned to the supplier and so do not enter the waste stream. The principal activities involving use of radioactive substances, and the waste they generate, are described in Box.

The waste produced by health-care and research activities involving radionuclides, and related activities such as equipment maintenance, storage, etc., can be classified as follows:

- sealed sources;
- spent radionuclide generators;
- low-level solid waste, e.g. absorbent paper, swabs, glassware, syringes,vials;
- residues from shipments of radioactive material and unwanted solutions of radionuclides intended for diagnostic or therapeutic use;
- liquid immiscible with water, such as liquid scintillation-counting residues used in radioimmunoassay, and contaminated pump oil;
- waste from spills and from decontamination of radioactive spills;
- excreta from patients treated or tested with unsealed radionuclides;
- low-level liquid waste, e.g. from washing apparatus;
- gases and exhausts from stores and fume cupboards.

Sources of Health Care Waste

The sources of health-care waste can be classed as major or minor according to the quantities produced.

While minor and scattered sources may produce some health-care waste in categories similar to hospital waste, their composition will be different. For example:

- they rarely produce radioactive or cytostatic waste;

- human body parts are generally not included;

- sharps consist mainly of hypodermic needles.

Health Care and Research Involving Radionuclides, and Waste Produced

Nuclear Medicine Laboratories

Unsealed source

Diagnostic procedures (organ imaging, tumour localization): use preparations with activities up to 800MBq (or even 6000MBq for certain lung-imaging techniques) and short half-life. Over 90% of diagnostic nuclear medicine applications use 99mTc.

Therapeutic applications (radiotherapy): use preparations of ^{32}P, ^{125}I and ^{131}I, which are of a much higher level of activity. However, these applications are infrequent. They are used in the activity range of up to 1GBq to treat hyperthyroidism and up to 10GBq to treat thyroid carcinoma.

Generated waste: glassware, syringes, absorbent paper, solutions, excreta from patients treated or tested with unsealed radionuclides. Waste from diagnostic procedures is usually low-level; wastes from therapeutic applications, however, may be relatively high-level. All radionuclides used have relatively short half-lives (between 6 hours and 60 days).

Sealed Sources

Therapeutic applications: use sealed sources that generally involve radionuclides with high activity levels and long half-lives (e.g. cobalt, caesium). In teletherapy the source is comparatively distant from the patient's body; brachytherapy usually employs small sources to deliver doses at distances up to a few centimetres, by surface, intracavitary, or interstitial application.

Generated waste: these activities do not routinely generate radioactive waste. Sources should be reused as long as is feasible, or returned to the supplier when exhausted or no longer required.

Research Laboratories

Generated waste: significant quantities of ^{14}C and ^{3}H (both with long half-lives) are used in research activities, which therefore generate large volumes of waste with low activity.

Clinical Laboratories

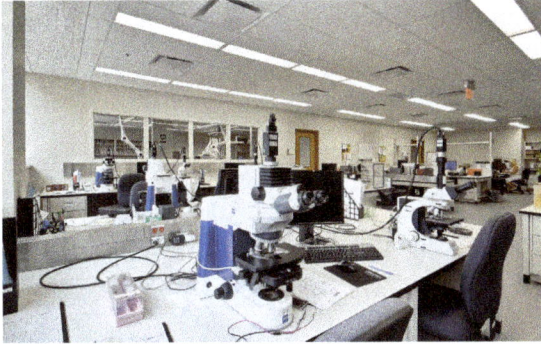

Generated waste: laboratories involved in radioimmunoassay produce relatively large volumes of waste with low radioactivity, including gases (e.g. ^{85}Kr, ^{133}Xe).

The composition of wastes is often characteristic of the type of source. For example, the different units within a hospital would generate waste with the following characteristics:

1. Medical wards: mainly infectious waste such as dressings, bandages, sticking plaster, gloves, disposable medical items, used hypodermic needles and intravenous sets, body fluids and excreta, contaminated packaging, and meal scraps.

2. Operating theatres and surgical wards: mainly anatomical waste such as tissues, organs, fetuses, and body parts, other infectious waste, and sharps.

Major sources of Health-care Waste

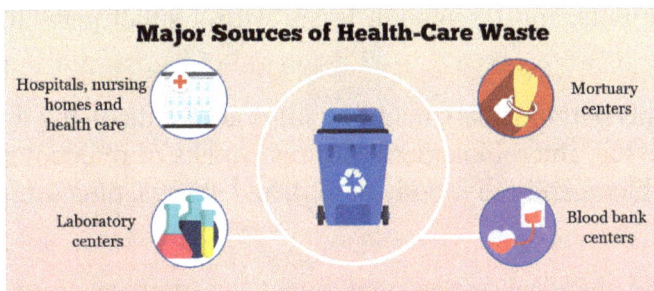

Hospitals

— University hospital

— General hospital

— District hospital

Other health-care establishments

— Emergency medical care services

— Health-care centres and dispensaries

— Obstetric and maternity clinics

— Outpatient clinics

— Dialysis centres

— First-aid posts and sick bays

— Long-term health-care establishments and hospices

— Transfusion centres

— Military medical services

Related laboratories and research centres

— Medical and biomedical laboratories

— Biotechnology laboratories and institutions

— Medical research centres

Mortuary and autopsy centres

Animal research and testing

Blood banks and blood collection services

Nursing homes for the elderly

Other health-care units: mostly general waste with a small percentage of infectious waste.

Laboratories: mainly pathological (including some anatomical), highly infectious waste (small pieces of tissue, microbiological cultures, stocks of infectious agents, infected animal carcasses, blood and other body fluids), and sharps, plus some radioactive and chemical waste.

Pharmaceutical and chemical stores: small quantities of pharmaceutical and chemical wastes, mainly packaging (containing only residues if stores are well managed), and general waste.

Support units: general waste only.

Health-care waste from scattered sources generally has the following characteristic composition:

• Health care provided by nurses: mainly infectious waste and many sharps.

• Physicians' offices: mainly infectious waste and some sharps.

- Dental clinics and dentists' offices: mainly infectious waste and sharps, and wastes with high heavy-metal content.

- Home health care (e.g. dialysis, insulin injections): mainly infectious waste and sharps.

Biological Waste

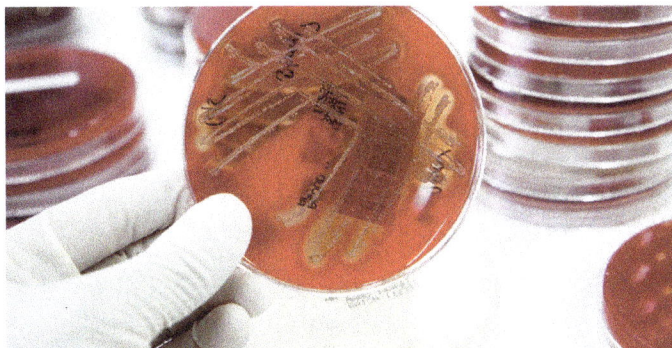

Biological waste must be managed separately from chemical waste. The most common example where chemical waste is mistaken for biological waste is agarose gel contaminated with ethidium bromide or heavy metals (i.e. arsenic, chromium). This type of material should always be managed as chemical waste. When both chemical and biological waste types exist, the biological agents should be treated first.

The following sections offer general safety guidelines and procedures for disposing of biological waste.

Segregation

- Segregation is necessary when working with hazardous biological agents.

- Any waste that could cause a laceration or puncture must be disposed of as "Sharps." Sharps must be segregated from other waste.

- Do not mix waste that requires incineration with glass or plastics.

- Do not mix biological waste with chemical waste or other laboratory trash.

- Segregate hazardous biological waste from non-hazardous biological waste.

Handling and Transport

Follow these guidelines for handling and transporting biohazardous waste:

- Properly trained personnel (not the custodial staff) are responsible for transporting treated biological waste to the dumpster or incinerator. Only properly trained technical personnel may handle untreated biohazardous waste.

- Contain and label all treated waste before transporting it to the incinerator or dumpster.

- Avoid transporting untreated biohazardous materials and foul or visually offensive materials through nonlaboratory areas.

- Do not use trash/laundry chutes, compactors, or grinders to transfer or process untreated biohazardous waste.

Labeling Biohazardous Waste

Follow these guidelines for labeling biohazardous waste:

- Clearly label each container of untreated biohazardous waste and mark it with the Biohazard Symbol.

- Label containers intended for landfill disposal to indicate the method of treatment. Cover the Biohazard Symbol with this label.

- Label autoclave bags with special tape that produces the word "Autoclaved" upon adequate thermal treatment. Apply this tape across the Biohazard Symbol before autoclaving the bag.

- Label all containers for sharps as "Encapsulated Sharps."

- It is recommended to label nonhazardous biological waste as "Nonhazardous Biological Waste".

Disposal Methods

Different materials require different disposal methods to ensure safety. Follow these guidelines for physically disposing of biological waste.

- Animal Carcasses and Body Parts:

  Incinerate the materials or send them to a commercial rendering plant for disposal.

- Solid Animal Waste:

  All animal waste and bedding that is infectious or harmful to human, animals, or the environment should be treated by incineration, thermal disinfection, or chemical disinfection.

- Liquid Waste:

  Liquid waste, including bulk blood and blood products, cultures and stocks of etiological agents and viruses, cell culture material, and rDNA products should be disinfected by thermal or chemical treatment and then discharged into the sanitary sewer system.

- Metal Sharps:

  All materials that could cause cuts or punctures must be contained, encapsulated, and disposed of in a manner that does not endanger other workers. Needles, blades, etc. are considered biohazardous even if they are sterile, capped, and in the original container.

- Pasteur Pipets and Broken Glassware:

  Place in a rigid, puncture resistant container. Disinfect by thermal or chemical treatment, if contaminated. Label the container as "Broken Glass" and place it in a dumpster.

NOTE

If broken glass is commingled with metal sharps, encapsulation is required for disposal.

- Plastic Waste:

  Contaminated materials must be thermally or chemically treated and placed in a properly labeled, leak-proof container for disposition in the dumpster. Materials that are not contaminated may be placed directly in the dumpster.

- Microbiological Waste:

  Solids must be thermally or chemically treated and placed in a properly labeled, leak-proof container for disposition in the dumpster. Liquids must be thermally or chemically treated and then discharged into the sanitary sewer system.

- Human Pathological Waste:

Human cadavers and recognizable body parts must be cremated or buried. Other pathological waste from humans and primates must be incinerated.

- Genetic Material:

Materials containing rDNA or genetically altered organisms must be disposed of in accordance with NIH Guidelines and the Texas State University Waste Disposal Program.

Non-hazardous Biological Waste

Most biological waste that is not infectious or otherwise hazardous to humans, animals, plants, or the environment may be discarded as regular waste or sewage. The only exceptions are animal carcasses and body parts. These wastes must be incinerated or sent to a commercial rendering plant for treatment. In addition, there are no record-keeping requirements for non-hazardous biological waste.

Follow these guidelines for non-hazardous biological waste:

- It is recommended to autoclave or disinfects all microbial products, even if they are not biohazardous.

- Avoid disposing of waste in a manner that could cause visual or odorous problems.

- Do not label non-hazardous biological waste as hazardous (e.g., do not use the Biohazard Symbol, red bags, etc.). Instead, it is recommended to label the container as "Nonhazardous Biological Waste".

- Use nonhazardous animal bedding and manure for compost or fertilizer when possible.

Clinical Waste

Clinical waste is any waste which poses a threat of infection to humans. The term also includes drugs or other pharmaceutical products.

Clinical waste is mainly produced by hospitals, health clinics, doctors' surgeries and veterinary practices, but also arises from private households.

Examples of clinical waste include:

- Human or animal tissue.

- Blood or other bodily fluids.

- Excretions.

- Drugs or other pharmaceutical products.

- Swabs or dressings.

- Syringes, needles or other sharp instruments.

Clinical Waste Types

Offensive Waste

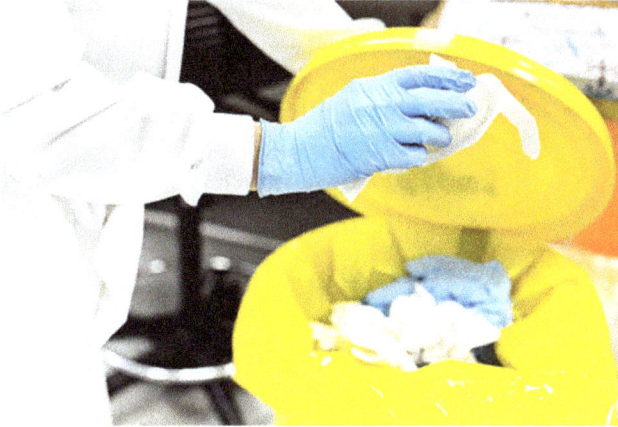

Offensive waste is waste generated by healthcare activities that are not infectious and don't contain medicines, chemicals or sharps, but may cause offence to anyone who comes into contact with it. It's a legal requirement to segregate hazardous wastes (such as infectious waste) from non-hazardous wastes (such as offensive waste) to ensure that segregation is maintained and mixing is prevented.

Offensive Waste is classified under (EWC 18-01-04) and is defined in The Controlled Waste Regulations 1992 some waste types that fall into this category include, but are not limited to;

- Sanitary hygiene waste

- Incontinence pads

- Nappy waste

- Some tattooist wastes

- Some dental waste

- Gloves, gowns and masks and other outer protective garments or dressings that have not been contaminated with bodily fluids

This waste stream is usually disposed of via incineration or taken to a non-hazardous landfill site by a registered waste carrier.

Offensive Waste is generally disposed of into internal clinical bins and yellow tiger stripe bags, plus external wheelie bins from 240 Litres up to 1000 Litres.

External wheelie bins are emptied on agreed frequencies with wheelie bin package collections being the most cost effective method of disposing of your waste.

Infectious Waste

Infectious waste mostly classified as (EWC 18-01-03) is also referred to as Hazardous Clinical waste and is mainly produced by but not limited to healthcare establishments such as hospitals, clinics, doctor's surgeries, dental, nursing homes and veterinary practises.

Infectious waste is waste that contains viable micro-organisms or their toxins (known collectively as pathogens) which are known or reliably believed to cause disease in humans or other living organisms. Any waste that is known or reliably believed not to contain viable pathogens is not an infectious waste.

Infectious waste can be any kind of waste which contains properties which may be harmful to human health or the environment. It could be completely or partly made up from:

- Blood or other bodily fluids
- Swabs or dressings
- Human or animal tissue
- Drugs or pharmaceutical products
- Needles, syringes and other sharp implements
- Excretions

Sharps Waste

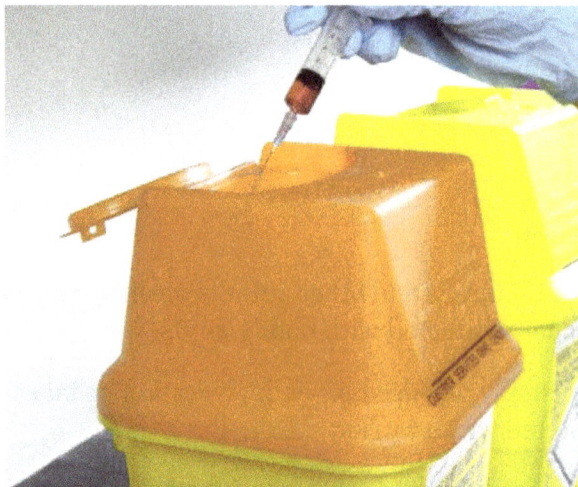

Sharps waste is a subset of infectious waste and comprises syringes, needles, lancets, broken glass and any other materials that can pierce the skin. The combination of contamination with pathogens and the ability to break through the skin's protection make them one of the most dangerous wastes produced in healthcare.

The vast majority of sharps waste is syringes from the 16 billion injections given each year. Vaccinations are essential to prevent disease, but over half of the curative injections are not necessary as they could be replaced by oral medications.

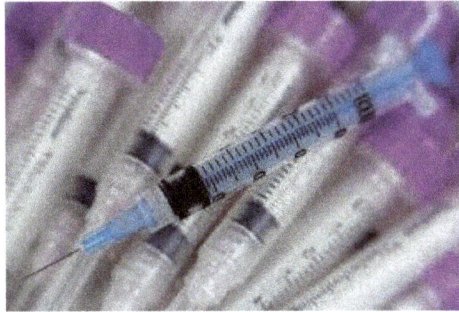

Reuse of syringes causes millions of infections each year, with HIV, hepatitis and bacterial infections. To try to reduce this, WHO recommends use of auto-disable syringes during vaccination programs. However, these can still cause injury and 10-20% of needle stick injuries happen during disposal, making proper management essential.

Moreover, there is a thriving trade in second hand syringes in several countries, notably in South Asia. These are repacked and sold to unwitting customers. Not only do they do untold harm to patients, but the rag-pickers who search them out get 3-5 needle stick injuries a day.

Needle cutters- also called hub cutters- which cut off the needle and the end of the syringe so that it cannot be used again, can prevent reuse and make treatment and disposal safer and easier.

Sharps can easily be autoclaved or disinfected with any of the technologies used for infectious waste. In addition, some waste treatment companies provide reusable sharps containers that can be disinfected with their contents and returned to use.

Reusable containers can also be made from aluminium by local metalworkers to suit the needs of each situation.

Finally, syringes are made from high grade plastic and they have been disinfected and made non-reusable via needle removal or shredding, they can often be recycled.

In addition to needles and blades, anything attached to them will also be considered sharps waste, such syringes and injection devices.

Blades can include razors, scalpels, X-Acto knife, scissors, or any other medical items used for cutting in the medical setting, regardless of if they have been contaminated with biohazardous material. While glass and sharp plastic are considered sharps waste, their handling methods can vary.

Glass items which have been contaminated with a biohazardous material will be treated with the same concern as needles and blades, even if unbroken. If glass is uncontaminated, it is still often treated as a sharp, because it can break during the disposal process. Contaminated plastic items which are not sharp can be disposed of in a biohazardous waste receptacle instead of a sharps container.

Dangers Involved

As a biohazardous material, injuries from sharps waste can pose a large public health concern. By penetrating the skin, it is possible for this waste to spread blood-borne pathogens. The spread of these pathogens is directly responsible for the transmission of blood-borne diseases, such as hepatitis B (HBV), hepatitis C (HCV), and HIV. Health care professionals expose themselves to the risk of transmission of these diseases when handling sharps waste. The large volume handled by health care professionals on a daily basis increases the chance that an injury may occur.

The general public can occasionally be at risk to injuries from sharps waste as well when improperly disposed of by injection drug users.

Sharps Containers

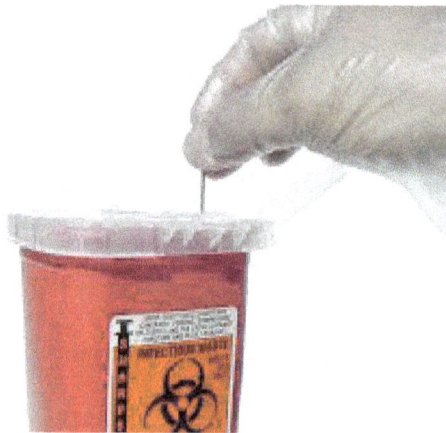

A sharps container is specially designed for safe disposal of sharps waste.

A sharps container is a hard plastic container that is used to safely dispose of hypodermic needles and other sharp medical instruments, such as an IV catheters and disposable scalpels. Sharps containers may be single use which are disposed of with the waste inside or reusable which are robotically emptied and sterilized before being returned for re-use.

Needles are dropped into the container through an opening in the top. Needles should never be pushed or forced into the container, as damage to the container and/or needle stick injuries may result. Sharps containers should not be filled above the indicated line, usually two-thirds full.

In North America, sharps containers are often red, and elsewhere are often yellow.

Airports and large institutions commonly have sharps containers available in restrooms for safe disposal for users of injection drugs, such as insulin-dependent diabetics. People injecting drugs in their homes may substitute other hard-sided containers such as empty milk jugs for disposal of needles.

Extreme care must be taken in the management and disposal of sharps waste. The goal in sharps waste management is to safely handle all materials until they can be properly disposed. The final step in the disposal of sharps waste is to dispose of them in an autoclave. A less common approach is to incinerate them; typically only chemotherapy sharps waste is incinerated. Steps must be taken along the way to minimize the risk of injury from this material, while maximizing the amount of sharps material disposed.

Health care workers are to minimize their interaction with sharps waste by disposing of it in a sealable container. Attempts by health care workers to disassemble sharps waste is kept to a minimum. Strict hospital protocols and government regulations ensure that hospital workers handle sharps waste safely and dispose of it effectively.

Self-locking and sealable sharps containers are made of plastic so that the sharps cannot easily penetrate through the sides. Such units are designed so that the whole container can be disposed of with other biohazardous waste. Single use sharps containers of various sizes are sold throughout the world. Large medical facilities may have their own 'mini' autoclave in which these sharps containers are disposed of with other medical wastes. This minimizes the distance the containers have to travel and the number of people to come in contact with the sharps waste. Smaller clinics or offices without such facilities are required by federal regulations to hire the services of a company that specializes in transporting and properly disposing of the hazardous wastes.

NIOSH found through results from focus groups that accommodation, functionality, accessibility, and visibility are four areas of high importance to be able to ensure safe discarding of sharps. The studies found it was important to have containers that are easy to use with little need for training to be able to use. The containers should be visible in any areas that sharps are used and be placed in such degree that spillage and injury will not be likely to occur with use.

Recent legislation in France has stated that pharmaceutical companies supplying self-injection medications are responsible for the disposal of spent needles. Previously popular needle clippers and caps are no longer acceptable as safety devices and either sharps box or needle destruction devices are required.

Disposal methods vary by country and locale, but common methods of disposal are either by truck service or, in the United States, by disposal of sharps through the mail. Truck service involves trained personnel collecting sharps waste, and often medical waste, at the point of generation and hauling it away by truck to a destruction facility. Similarly, the mail-back sharps disposal method allows generators to ship sharps waste to the disposal facility directly through the U.S. mail in specially designed and approved shipping containers. Mail-back sharps disposal allows waste generators to dispose of smaller amounts of sharps more economically than if they were to hire out a truck service.

Injection Technology

With more than sixteen billion injections administered annually worldwide, they are the largest contributor to sharps waste. For this reason many new technologies surrounding injections have been developed, mostly related to safety mechanisms. As these technologies have been developed governments have attempted to make them commonplace to ensure sharps waste safety. In 2000, the Needle stick Safety and Prevention Act was passed, along with the 2001 Blood borne Pathogens Standard.

Safety syringes help reduce occurrences of accidental needle sticks. One of the most recent developments has been the auto-disable injection device. These injection devices automatically disable after a single use. This can be done by retracting the needle back into the syringe or rendering the syringe plunger inoperable. With the injection device now inoperable, it cannot be reused. Shielding the needle after the injection is another approach for safe management of sharps. These are hands free methods usually involving a hinging cap that can be pressed on a table to seal the needle. Another technology in sharps waste management relating to injections is the needle remover. Varying approaches can be taken with the main goal to separate the needle from the syringe. This allows the sharp needle to be quarantined and disposed of separate from the syringe. There is debate around the use of these devices as they involved in additional step in the handling of sharps waste.

In the Developing World

Sharps waste is of great concern in developing and transitional regions of the world. Factors such as high disease prevalence and lack of health care professionals amplify the dangers involved with sharps waste, and the cost of newer disposal technology makes them unlikely to be used. As with the rest of the world injection wastes make up the largest portion of sharps waste. However, injection use is much more prevalent in this world segment. One of the contributors to this increase is a larger emphasis placed on injections for therapeutic purposes. It has been estimated that 95% of all injections in developing regions are for therapeutic purposes. The average person has been estimated to receive up to 8.5 injections per year. Newly developed injection technologies are rarely used to provide these injections due to added costs. Therefore, the majority of injections are given with standard disposable syringes in developing regions.

The infrastructure of developing regions is not equipped to deal with this large volume of contaminated sharps waste. Contrary to the industrialized world, disposal incinerators and transportation networks are not always available. Cost restraints make the purchase of single use disposable containers unrealistic. Facilities are often overwhelmed with patients and understaffed with educated workers. Demand on these facilities can limit the emphasis or enforcement of waste disposal protocols. These factors leave a dangerous quantity of sharps waste in the environment. Contrasts between the industrialized and developing world segment can be seen in accidental needle stick injuries.

Storage of Medical Waste

Labelling

These labels are warnings to employees and the public about the type of waste in the container. Your local jurisdiction may have rules about warning labels that must be affixed to storage containers. Even if there are no legal requirements, best practices call for warnings that indicate the nature of the hazard. For infectious, pathological, and most regulated medical waste, the medical waste symbol can be employed. Indeed if you buy bags and containers designed for medical waste, these often come with the medical waste symbol on it.

The hazardous waste symbol, although not as widely recognized as the medical waste symbol, is appropriate for RCRA waste.

Radioactive waste (often produced in healthcare facilities) and mixed waste (mix of radioactive and hazardous waste) should be labeled with the universal radiation symbol.

Canadian guidelines stipulate a special Cytotoxic warning symbol on cytotoxic drug waste. In the US, cytotoxic waste would be classified as RCRA waste.

Labelling for Internal use

Well-run facilities (if they are large), require labels on waste bags and containers for tracking purposes. It is not different from an inventory management system. Employees (either those who generate the waste or waste technicians) should be apply a label with the date, type of waste, and point of generation. Doctors, dental, and veterinary offices are usually small enough to not do this, but any hospital should. The weight of the waste should be recorded. This information will be useful in the long run for identifying problems in cross-contamination among waste streams and mistakes in segregation.

OSHA has requirements for worker safety when it comes to blood and "other potentially infectious material". These include labeling of storage containers with the the biohazard symbol, and the term "biohazard." The background of the label must be fluorescent orange or orange-red. OSHA does allow facilities to use red bags instead of biohazard labels.

Bags

The marketplace is full of vendors that sell bags for medical waste. Just look in a directory of industrial supplies or an internet search. For the most part there are no (or few) government regulations on these (containers for sharps are another matter.) The vendors may brag their bags are thick and resist punctures. Some bags are labeled with the biohazard symbol and some are colored. Some manufacturers sell both the trash can (specifically labeled for medical waste) and the plastic (or canvas) bag that goes with it. Many manufacturers make medical waste bads red, and some waste management plans specify red bags be used.

Best practice for most facilities is to buy these dedicated medical waste bags. Could you just use general purpose garbage bags? In a pinch you could and these will work for most medical waste items, provided you don't overfill them. But in general you want to keep medical waste in containers labeled medical waste, and it is worth paying a little more to get bags and bins already labeled.

Bags sold for medical waste are usually made from linear low-density polyethylene, although some may be made from polypropylene. High-density polyethylene is also used. One good thing about these materials is that they will burn in an incinerator with almost no ash. Bagged waste does not need to be removed from the bag before it enters the furnace. Some processes move the bags through a shredder first, which breaks the bags and may mix the waste from one bag with another.

Waste bags should be filled to no more than three quarters full. At that point they should be closed for collection. Don't staple the bag closed. Either use a twisty, or, if the bag comes ready to self-seal, employ that. Keep spare bags in areas where waste is collected in bags, so employees are not tempted to overfill any one bag.

Bags come in sizes from sandwich size to as big as 55-gal drums (a standard drum size in industry.) You might hear of "red bags" and "red bag waste". You might think this designation has a legal meaning, but it doesn't. There is no rule that says your infectious or pathological waste must be put in a red bag, although most industrial hygienists and sanitation engineers probably prefer a red bag, all other things being equal.

One disadvantage of bags shows up when autoclave treatment is used. To ensure the waste is heated adequately, the bags must be opened to allow the steam in. When bagged waste is to be incinerated, this is not a concern, as the bags themselves are combustible. However, some plastic bags may melt in the autoclave, producing a mess. To avoid this, test bags in the autoclave with no waste. Bag suppliers might be able to specify a temperature the plastic will stand up to.

The World Health Organization recommends infectious waste bags be a minimum of 70 μm in thickness (ISO 7765 2004). It is easy to find bags thicker than that.

Containers

More sturdy containers are employed to hold large quantities of waste. You can buy these from many retailers and there are no "official" requirements for them. If a legal dispute arises, the court will look at whether the waste manager exercised due care. Containers, bins, barrels for medical waste can be made from various materials (plastic, steel, aluminum). Colors should conform with whatever scheme you have set for your facility.

Large containers should be lockable, especially if they are going to be outdoors or in an area where a lot of customers or non-professionals frequent.

One mistake facility managers make is assuming that the container a material was delivered in when it is purchased is acceptable as a waste container for the same material.

Once the material becomes a waste, it is subject to different rules and regulations. The in-bound shipping container will probably not suffice as a storage or our-bound shipping container.

If you have a waste removal service, they might be able to provide containers. Considering the low cost of the containers compared to what they charge, they really should do so if they want you as a regular customer. They may also have containers that fit with their handtrucks or other mechanical transport means.

Rubbermaid makes these convenient containers with foot pedals. We do not endorse them (or any other products), but be aware that they are many potential containers.

Dumpsters

You don't want to put regulated waste in dumpsters, which are not built tight enough to keep out rain and vermin. Outdoor dumpsters are ok for municipal solid waste, but regulated waste (hazardous, radioactive, biological) should be kept in tight containers and indoors.

Sharps Containers

These containers are designed so the user is never exposed to any of the sharps already in the container, eliminating the possibility of contact or puncture by any of the used needles. Sharps containers are generally made of thick plastic, and have a door that opens and the user can insert the sharp into the container. When the door is closed, the sharp is dropped down into the main chamber of the container. The container functions much like a standard post office mailbox, in that the user cannot reach the sharps inside the container via the door. Sharps containers are also used for other categories of sharps, including scalpels and lancets.

Sharps containers are found in public locations, and can also be used by private individuals who use sharps in the home, such as diabetics who require regular injections of insulin. For home users, a sharps container is provided by a private disposal company. When the individual fills the container, the container is then mailed to the private disposal company for disinfection and destruction prior to disposal.

Keeping the Waste on Site

Storage rooms should have locks and be away from the public and most building occupants. Ideally they should be indoors, but any outdoor storage should be fenced to keep away animals and humans.

State regulations often dictate the maximum amount of time that medical waste can be stored prior to treatment. For example, in the state of New York, storage of regulated medical waste is limited to seven days. Medical waste must be stored separately from standard waste, without possibility of the two waste types mixing. All reusable storage containers must be disinfected after they have been emptied, unless they employ a disposable liner that is removed with the waste.

Government regulations require that certain materials be managed in specially designed biosafety level areas.

Note the workplace safety rules should always be followed with medical waste, and there are cases of workers getting sick, including with tuberculosis, from contact with waste.

The International Committee of the Red Cross (ICRC) recommends storage time for infectious waste not exceeds 72 hours in winter and 48 hours in summer in temperature climate zones. In hot climates, the recommended limits are 48 hours in the cooler season and 24 hours in the warmer season. If refrigerated storage is available, hold times can go up to a week, per the IRCR.

The ICRC recommends lists criteria for storage areas. While this does not have the force of law, they are good guidelines and should help you with regulators:

1. Area should be closed, and access must be restricted to authorized persons only.

2. Waste storage area must be separate from food store; Area should be covered and sheltered from the sun.

3. Flooring must be waterproof with good drainage.

4. Must be easy to clean.

5. Must be protected from rodents, birds and other animals; Must allow easy access for on-site and off-site means of transport.

6. Should be well lighted.

7. Should be big enough to sort waste if possible and to allow physical separation of different categories of waste. Ideally, it should have some physical barriers to prevent mixing of wastes of different categories.

8. Should be close to any on-site treatment.

9. Compartmented (so that the various types of waste can be sorted).

10. Should have eye-washed and other PPE.

11. Should be near wash basins.

12. The entrance must be marked with a sign to discourage people from entering unless they need to be there and warning of hazards.

The Difference between Storage and Disposal

The patent literature is rife with disposal containers for medical waste. But these are actually storage containers for temporary isolation of the waste from human workers. (Disposal refers to long-term and presumably permanent resting place for waste.) Storage is of concern to the waste manager, too, and vendors provide a wide variety of options for containers. Your state government still might have something to say about storage but the containers on the market usually are acceptable to them.

Operating procedures for your business should specify the maximum time you intend to store waste on site. If you know how fast your waste is generated and how often it gets taken off-site you will be able to determine the volume and type of storage needed.

Design of Storage Systems

Large operations often have small collection bins spread around the facility. For instance, a complex of doctors' offices may have containers in each examination room, with a regular schedule of transferring this waste to a larger storage unit.

The placement of the storage units is an element in the overall design of the waste management process. Factors that must be weighed:

1. Number of storage units at a given facility (fewer is better to reduce risk of release).

2. Distance between generation of waste and storage unit (less distance is better – many operating rooms have units so doctors and nurses can put tissue in units without moving long distances).

3. Size of storage units.

4. Air flows in locations.

5. Ergonomics.

6. Traffic patterns (people, vehicles) nearby.

7. When storing liquid chemicals, the storage area should have a sump for spill collection..

8. Odor control.

Treatment of Medical Waste

The treated waste - if sufficiently sterile - can generally be disposed with waste in a sanitary landfill, or in some cases discharged into the sewer system. In the past, treatment of medical waste was primarily performed on-site at hospitals in dedicated medical waste facilities. Over time, the expense and regulation of these facilities have prompted organizations to hire contractors to collect, treat, and dispose of medical waste, and the percentage of medical organizations that perform their own treatment and disposal is expected to drop.

To ensure that each treatment method provides the proper environment for the destruction of biologicals, test indicators for microbiological spores measure the treatment effectiveness. Microbiological spores are the most difficult of biologicals to destroy, so when the test package cannot be cultured after treatment, the waste is considered properly treated. In treatment methods where shredding or maceration is employed, the test package is inserted into the system after the shredding process to avoid physical destruction of the test package. The test package is then retrieved from the waste after treatment.

Incineration

Incineration is the controlled burning of the medical waste in a dedicated medical waste incinerator. Among industry professionals, these units are often referred to as hospital/medical/infectious waste incinerators (HMIWIs).

Incineration is an old technology and was widely used in the past for all sorts of waste. Individual buildings had their own waste incinerators in many cases. Incinerators got a bad reputation because of the air pollution they created and because the bottom ash, or sinter, was hard to keep under control. Lay members of the public unfortunately still have negative associations with incinerators. Today's incineration units are typically much cleaner.

There are parts of the world where open pit burning still take place. And accidental fire - e.g. a house on fire - produce flames and smoke and debris. This makes "burning" and "combustion" bad words, but when approached from a cold hard engineering standpoint, incineration is often the best technology for treating medical waste. It can eliminate pathogens - even hard-to-kill bacterial spores - and can reduce the volume and mass of waste that goes to landfills considerably. Incineration can break down and render harmless hazardous organic chemicals. With proper technology, little acid gas is released to the atmosphere.

There are fewer medical waste incinerators operating in the United States today than there were decades ago when practically every hospital had an incinerator. Between 1998 and 2008 the number of hospital incinerators fell by more than 95 percent. However, incineration technology is still an important part of the medical waste management landscape. Central incineration facilities that take waste from different hospitals are more common.

Because most medical waste can be incinerated, the waste is not sorted or separated prior to treatment. Incineration has the benefit of reducing the volume of the waste by 80 percent or more, sterilizing the waste, and eliminating the need for pre-processing the waste before treatment. The resulting incinerated waste can be disposed of in traditional methods, such as brought to a landfill. The downside of incineration is potential pollution from emissions generated during incineration. The EPA has stringent requirements on emissions from medical incinerators. The incineration process can be applied to almost all medical waste types, including pathological waste, and the process reduces the volume of the waste by up to 90%.

Modern incinerators incinerator can provide a secondary benefit by creating heat to power boilers in the facility.

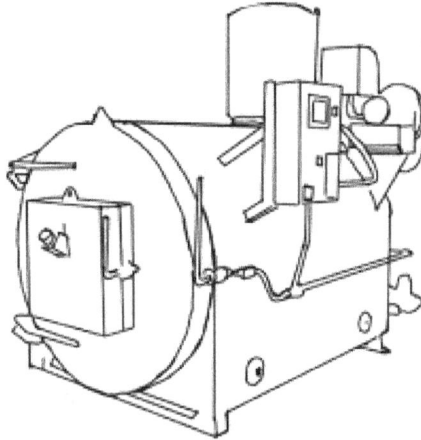

The largest concern associated with incineration is air pollution from emissions. The EPA says that at least 20% of medical waste is plastic. The biggest concern is that the incinerator may create toxic compounds. Dioxins and furans can be produced when these plastics burn. Older medical waste incinerators included no pollution control equipment. As new federal and state emission regulations are instituted that have more stringent requirements, medical incinerators are often not being replaced at the end of their service life. Over time, the amount of waste being incinerated will be reduced as other technologies replace on-site incinerators.

Another concern is incinerator ash. As incinerators are designed or retrofitted with pollution prevention equipment, more of the potentially toxic chemicals that previously ended up in emissions now remain in the ash. Incinerator ash is generally disposed of in landfills.

Downsides of Incinerators

The public often has an aversion to incinerators and may raise objections if they hear one is being put in their area. The popular perception of incineration is informed by pictures of open pit burning done decades ago in the US and still today in some countries. Open pit burning is indeed not effective enough and results in smoke and other undesirable materials being released to the atmosphere. Most people don't understand how incineration units can be made clean-burning and engineered to reduce the risk of dangerous releases. Most people also don't understand how many incinerators are in their area already.

However, incineration can be a dirty process if not controlled adequately or if the process has not been designed correctly. Incineration can produce.

- Fine particles (in the smoke): The particles can include heavy metals. If removed from the smoke before release, these particles are called fly ash and constitute another disposal problem.

- Acid gases: these are formed during burning. Chlorine compounds, when burned, yield hydrochloric acid. Sulfur compounds yield sulfur dioxide or sulfur trioxide. Nitrogen oxides are produced in any high temperature treatment.

- Ozone: indirectly. Nitrogen oxides from the exhaust can subsequently react with hydrocarbons in the air to produce ozone.

- Bottom ash or sinter: the ashes after the incineration process is complete. Mostly inorganic material. This is disposed of in a landfill - either sanitary one or a hazardous waste one.

- Heat: While good incinerators are insulated to save energy and protect workers, heat generation must be accounted for in facility and process design.

Ocean Incineration

Ship-mounted incinerators will burn your waste for a fee. The advantages of burning waste at sea is that it is away from population centers and, if far enough from land, is free from national laws and regulations. That means the incinerators can be less efficient and pollute more. Ash from the incinerators is often dropped into the ocean.

There may be international treaties about these operations - such regulations are beyond our scope. The waste manager may have ethical problems with these solutions as they seem an "easy-out" and way to skirt the rules and to be irresponsible.

Mobile Incinerators

Mobile incinerators tend to be small and have the same downsides as ship-bourne incinerator. Getting a mobile incinerator to pass EPA regulations can be a challenge. They are usually less efficient and often more prone to creating furans and dioxins in the exhaust.

Small-scale Incinerators

Small-scale incinerators were once common in the United States and are still used in much of the world. They are generally less efficient and more prone to releasing pollution than large, well-engineered incineration systems. While small incinerators are almost always superior to dumping medical waste with no treatment, they should be considered only if there is no better solution available.

Disposal of Medical Waste

There are mainly five technology options available for the treatment of bio-medical waste. They can be grouped as follows:

- Chemical processes
- Thermal processes

- Mechanical processes
- Irradiation processes
- Biological processes

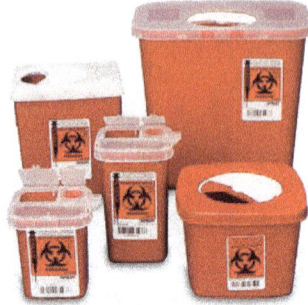

Chemical Processes

These processes use chemicals that act as disinfectants. Sodium hypochlorite, dissolved chlorine dioxide, peracetic acid, hydrogen peroxide, dry inorganic chemical and ozone are examples of such chemicals. Most chemical processes are water-intensive and require neutralising agents.

Thermal Processes

These processes utilise heat to disinfect. Depending on the temperature they operate, it is been grouped into two categories, which are Low-heat systems and High-heat systems

Low-heat systems (operates between 93 -177oC) use steam, hot water, or electromagnetic radiation to heat and decontaminate the waste. Autoclave & Microwave are low heat systems:

i. Autoclaving is a low heat thermal process and it uses steam for disinfection of waste. Autoclaves are of two types depending on the method they use for removal of air pockets. They are gravity flow autoclave and vacuum autoclave.

ii. Microwaving is a process which disinfects the waste by moist heat and steam generated by microwave energy.

High-heat systems employ combustion and high temperature plasma to decontaminate and destroy the waste. Incinerator & Hydroclaving are high heat systems.

Mechanical Processes

These processes are used to change the physical form or characteristics of the waste either to facilitate waste handling or to process the waste in conjunction with other treatment steps. The two primary mechanical processes are:

- Compaction - used to reduce the volume of the waste.

- Shredding - used to destroy plastic and paper waste to prevent their reuse. Only the disinfected waste can be used in a shredder.

Irradiation Processes

In these processes, wastes are exposed to ultraviolet or ionizing radiation in an enclosed chamber. These systems require post shredding to render the waste unrecognizable.

Biological Processes

Biological enzymes are used for treating medical waste. It is claimed that biological reactions will not only decontaminate the waste but also cause the destruction of all the organic constituents so that only plastics, glass, and other inert will remain in the residues.

Points to Ponder in Processing the Waste

Incineration

- Incinerators should be suitably designed to achieve the emission limits.

- Wastes to be incinerated shall not be chemically treated with any chlorinated disinfectants.

- Toxic metals in the incineration ash shall be limited within the regulatory quantities.

- Only low sulphur fuel like diesel shall be used as fuel in the incinerator.

Autoclaving

- The autoclave should be dedicated for the purpose of disinfecting and treating biomedical waste.

- When operating a gravity flow autoclave, medical waste shall be subjected to:

  ○ A temperature of not less than 121° and pressure of about 15 pounds per square inch (psi) for an autoclave residence time of not less than 60 minutes; or

  ○ A temperature of not less than 135 °C and a pressure of 31 psi for an autoclave residence time of not less than 45 minutes; or

  ○ A temperature of not less than 149 °C and a pressure of 52 psi for an autoclave residence time of not less than 30 minutes.

- When operating a vacuum autoclave, medical waste shall be subjected to a minimum of one per vacuum pulse to purge the autoclave of all air. The waste shall be subjected to the following :

  ○ A temperature of not less than 121°C and a pressure of 15 psi per an autoclave residence time of not less than 45 minutes; or

  ○ temperature of not less than 135 °C and a pressure of 31 psi for an autoclave residence time of not less than 30 minutes; or Medical waste shall not be considered properly treated unless the time, temperature and pressure indicate stipulated limits. If for any reason, these were not reached, the entire load of medical waste must be autoclaved again until the proper temperature, pressure and residence time were achieved.

Microwaving

- Microwave treatment shall not be used for cytotoxic, hazardous or radioactive wastes, contaminated animal carcasses, body parts and large metal items.

- The microwave system shall comply with the efficacy tests/routine tests.
- The microwave should completely and consistently kill bacteria and other pathogenic organism that is ensured by the approved biological indicator at the maximum design capacity of each microwave unit.

Deep Burial

- A pit or trench should be dug about 2 m deep. It should be half filled with waste, and then covered with lime within 50 cm of the surface, before filling the rest of the pit with soil.
- It must be ensured that animals do not have access to burial sites.
- Covers of galvanised iron/wire meshes may be used.
- On each occasion, when wastes are added to the pit, a layer of 10cm of soil be added to cover the wastes.
- Burial must be performed under close and dedicated supervision.
- The site should be relatively impermeable and no shallow well should be close to the site.
- The pits should be distant from habitation, and sited so as to ensure that no contamination occurs of any surface water or ground water.
- The area should not be prone to flooding or erosion.
- The location of the site will be authorized by the prescribed authority.
- The institution shall maintain a record of all pits for deep burial.

Disposal of Sharp Materials

- Blades and needles waste after disinfection should be disposed in circular or rectangular pits.
- Such pits can be dug and lined with brick, masonry, or concrete rings.
- The pit should be covered with a heavy concrete slab, which is penetrated by a galvanized steel pipe projecting about 1.5 m above the slab, within internal diameter of upto 20 mm.
- When the pipe is full it can be sealed completely after another has been prepared.

Radioactive Waste from Medical Establishments

- It may be stored under carefully controlled conditions until the level of radioactivity is so low that they may be treated as other waste.

- Special care is needed when old equipment containing radioactive source is being discarded.

- Expert advice should be taken into account.

Mercury Control

Wastes containing Mercury due to breakage of thermometer and other measuring equipment need to be given attention.

- Proper attention should be given to the collection of the spilled mercury, its storage and sending of the same back to the manufacturers.

- Must take all measures to ensure that the spilled mercury does not become part of biomedical wastes

- Waste containing equal to or more than 50 ppm of mercury is a hazardous waste and the concerned generators of the wastes including the health care units are required to dispose the waste as per the norms.

Waste Minimization

Waste minimization is an important first step in managing wastes safely, responsibly and in a cost effective manner. This management step makes use of reducing, reusing and recycling principles.

References

- Needle Remover Harner, C. (2004, October). Needle Remover Device Design Transfer Package. Retrieved September 7, 2005.

- Biowaste, waste: purdue.edu, Retrieved 24 April 2018

- Hazardous-and-clinical-waste, refuse-and-recycling, environment-and-waste: haringey.gov.uk, Retrieved 14 May 2018

- Waste-types: principalhygiene.co.uk, Retrieved 11 July 2018

- Treating-sharps-waste, global: noharm-global.org, Retrieved 28 March 2018

- Treatment: malsparo.com, Retrieved 18 March 2018

- Kotwal, Atul (March 2005). "Innovation, diffusion and safety of a medical technology: a review of the literature on injection practices". Social Science & Medicine. 60 (5): 1133–1147. doi:10.1016/j.socscimed.2004.06.044.

Chapter 4
Oil Refinery Waste Treatment

Oil refinery is an industrial plant concerned with the refinement of crude oil into useful products like gasoline, diesel fuel, kerosene, LPG, etc. It produces a considerable amount of industrial waste. All the diverse aspects of oil refinery waste treatment such as treatment of semi-processed oil refinery products, remediation of hydrocarbon contaminated soils, rehabilitation of storage tanks, etc. have been covered extensively in this chapter.

Oil Refinery Waste

Based on source and chemical composition the waste generated in refineries can be broadly classified as follows:

- Hydrocarbon Wastes: It includes API separator sludge, dissolved air floatation float, slop oil emission solids, tank bottoms, FFU sludge, desalter bottoms and waste oils/solvents.

- Spent Catalysts: It includes fluid cracking catalyst, hydro-processing catalyst and other spent inorganic clays.

- Chemical/Inorganic Wastes: It includes spent caustic, spent acids and waste amines.

- Contaminated Soils and Solids: It includes heat exchanger bundle cleaning sludge, waste coke/carbon/charcoal, waste sulphur and miscellaneous contaminated soils.

- Aqueous Waste: It includes biomass, oil contaminated water (not wastewater), high/low pH water and spent sulphide solutions.

Oily Sludge Source

Both the upstream and downstream operations in petroleum industry can generate a large amount of oily wastes. The upstream operation includes the processes of extracting, transporting, and storing crude oil, while the downstream operation refers to crude oil refining processes. In the upstream operation, the related oily sludge sources include slop oil at oil wells, crude oil tank bottom sediments, and drilling mud residues. A variety of oily sludge sources exist in downstream operation, including (a) slop oil emulsion solids; (b) heat exchange bundle cleaning sludge; (c) residues from oil/water separator, such as the American Petroleum Institute (API) separator, parallel plate interceptor, and corrugated plate interceptor (CPI); (d) sediments at the bottom of rail, truck, or storage tanks; (e) sludge from flocculation-flotation unit (FFU), dissolved air flotation (DAF), or induced air flotation (IAF) units, and (f) excess activated sludge from on-site wastewater biological treatment plant. Prior to being refined to petroleum products, crude oil is temporarily housed in storage tanks, where it has a propensity to separate into heavier and lighter petroleum hydrocarbons (PHCs). The heavier PHCs often settle along with solid particles and water. This mixture of oil, solids, and water deposited at the storage tank bottom is known as oily sludge. It is removed during tank cleaning operations and sent for further treatment or disposal.

The sludge quantity generated from petroleum refining processes depends on several factors such as crude oil properties (e.g., density and viscosity), refinery processing scheme, oil storage method, and most importantly, the refining capacity. Generally, a higher refining capacity is associated with a larger amount of oily sludge production. It has been estimated that one ton of oily sludge waste is generated for every 500 tons of crude oil processed16. It is estimated that more than 60 million tons of oily sludge can be produced every year and more than 1 billion tons of oily sludge has been accumulated worldwide. It is also expected that the total oily sludge production amount is still increasing as a result of the ascending demand on refined petroleum products worldwide.

The Formation of Crude Oil Sludge

Most of the oils have a property to separate into the heavier and lighter hydrocarbons during their storage and transportation. The heavy ends of crude oil are deposited on the bottom of storage tank.

Paraffin based crude oil sludge forms, when the molecular orbital of individual straight chain hydrocarbons are blended by proximity, producing an induced dipole force that resists separation. As the heavier straight chain hydrocarbons flocculate (heavier meaning predominantly the C_{20+} hydrocarbon molecules), they tend to fall out of suspension within a static fluid, as in the case of storage tanks, where they accumulate at the bot-

tom as viscous gel commonly known as sludge or wax. This newly formed profile strat-
ifies over time as the volatile components within the sludge are expelled with changes
in temperature and pressure. The departure of such volatile components results in a
concentrated heavier fractions within the sludge, accompanying with increased in density and viscosity, and decreased fluidity.

Asphaltic sludge is formed due to the formation of the tendency of asphaltenes, resins
and polymeric compounds to precipitate. Neutral resins are high molecular weight aromatic hydrocarbons. These resins are insoluble in alkalies and acids but are completely
miscible with petroleum oils, including light fractions.

Asphaltenes are similar to the neutral resins but they are insoluble in light gasoline and
petroleum ether. Asphaltenes are precipitated in presence of an excess of petroleum
ether. Asphaltenes and neutral resins are soluble in benzene, chloroform and carbon
disulphide.

Inorganic salts, sediments, sands, scale and dust are also present in the sludge.

Sludge Treatment Methods

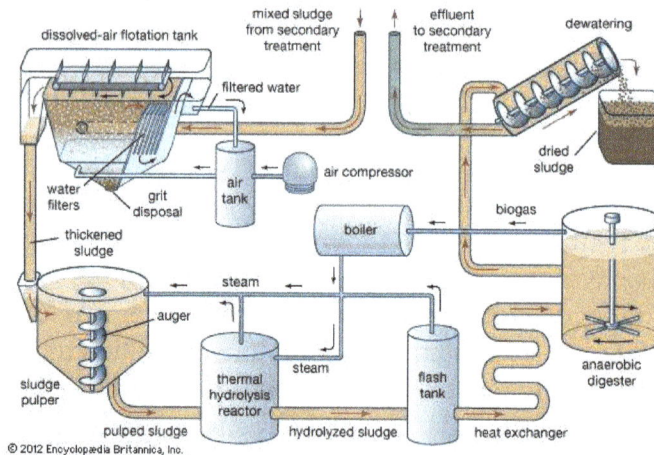

The various technologies for oil recovery and redemption of the crude sludge include
chemical treatments, various distillation processes, cracking, hydro-treating, solvent
treatment and bioremediation. Some of the conventional methods of sludge treatment
are as follows:

1. Manual Cleaning and Incineration

Manual cleaning is the low cost method. The cleaning is done by entering in the tank.
The sludge is moved out of the tank manually or to pumps present in the tanks. This
method takes long time. It is difficult to recover the usable hydrocarbons from the
sludge by using this method.

2. Solvent Extraction Method

Various solvents are used in this method, which are able to break down complex molecules present in the sludge into their basic constituents - water, crude oil and particulate. This method requires mixing and agitation apparatus. Sludge has waxy and non-waxy (asphaltenes) organic components along with salt, oxides and other inorganic materials. These may be dissolved by selecting appropriate solvent.

Oily sludge waste is firstly mixed in the reactor column with a solvent, which selectively dissolves the oil fraction of sludge and leaves the less soluble impurities at the column bottom. The oil solvent solution is then transferred to a solvent distillation system where the solvent is separated from oil. The separated oil is considered as oil recovery, while the separated solvent vapour can be liquefied through a compressor and cooling system and sent to a solvent recycling tank. The solvent can be used for repeating the extraction cycle. The bottom impurities from reactor column are pumped to a second distillation system, and the solvent contained in the impurities is separated and then sent to the solvent recycling tank, while the waste residues after separation may need further treatment.

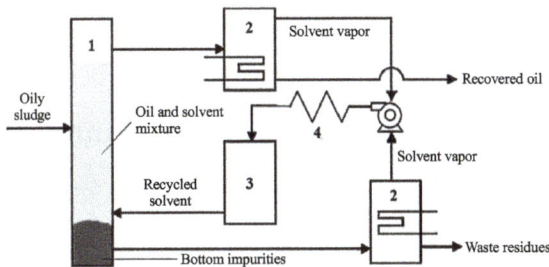

Fig: Block flow diagram of solvent extraction method

1. Reactor column; 2. Distillation; 3. Solvent recycling tank; 4. Compressor and cooling system.

Ultra High Temperature Gasification

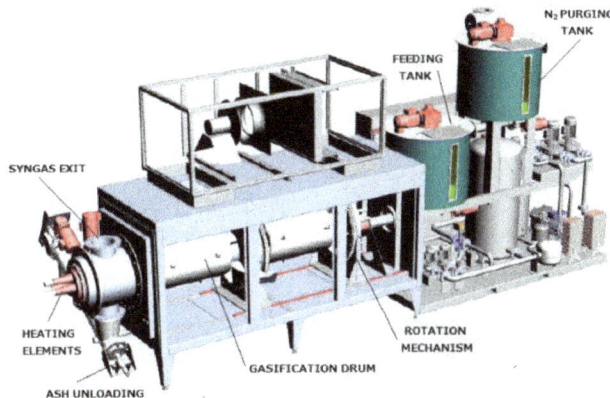

In this method, thermal oxidation of sludge is carried out. The sludge is heated to a very high temperature (10000 C) using plasma arc without oxygen. The sludge is converted to pyrogas by this method and this can be used as fuel.

Oil Sludge Separation using Cyclone

By this method, oil is recovered from the oily sludge and residue is separated.

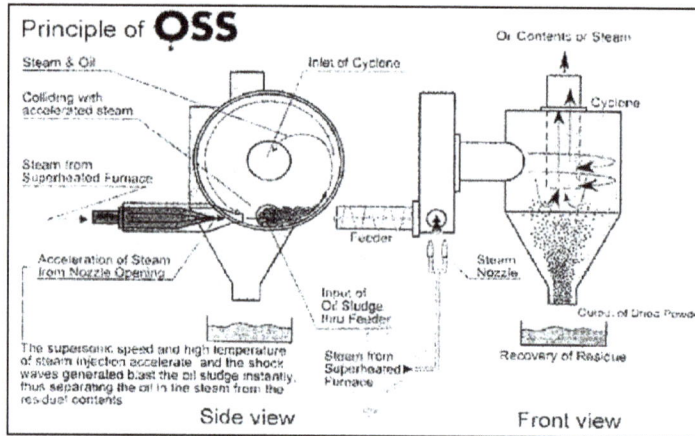

Fig: Oily sludge separation by cyclone

Oily Sludge Treatment by Application of Thermochemistry

In this method, oily sludge is diluted by water and heated, and then certain chemicals are added for extraction of oil from solid phase.

Microwave Heating Method

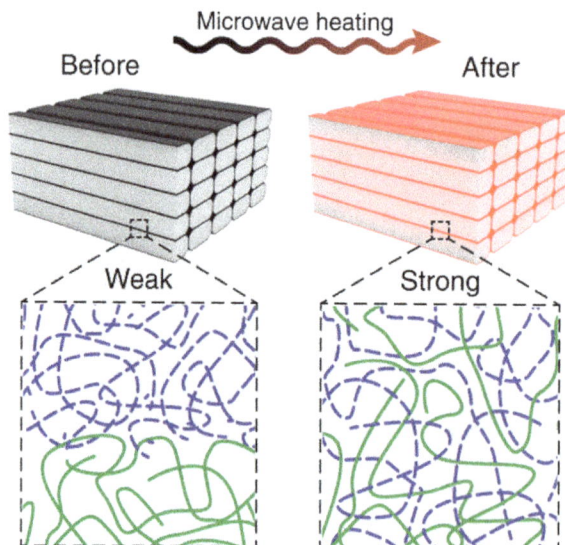

Microwave heating has more advantages than conventional heating. This method is used for waste management. Microwave heating technology for demulsification of sludge oil was first introduced by Klaila and Wolf. In conventional thermal heating, heat is transferred to the material through convection, conduction, and radiation of heat from the surfaces of the material. Whereas microwave energy is reached directly to the materials through molecular interaction with the electromagnetic field. In heat transfer, energy is transferred due to thermal gradients, but microwave heating is the transfer of electromagnetic energy to thermal energy and is energy conversion, rather than heat transfer. This results in many advantages for using microwaves for processing of materials. Microwaves can penetrate materials and deposit energy so heat can be generated throughout the volume of the material. The transfer of heat energy does not rely on diffusion of heat from the surface and it is possible to achieve rapid and uniform heating of thick materials.

Centrifugation Method

In this method, components are separated on the basis of their densities (such as water, solids, oil and pasty mixtures in oily sludge) by generating centrifugal force. This method uses a special high speed rotation equipment by reducing viscosity of oily sludge by adding organic solvents, demulsifying agents & tensioactive chemicals and the injection of steam and direct heating. The performance of centrifugation method is enhanced and energy consumption is reduced. Conway31 reported that after viscosity reduction using heat pre-treatment, the less viscous petroleum sludge could be effectively treated by a disc/bowl centrifuge, with more than 80% of the waste volume being obtained as liquid effluent from the first centrifugation, and residue from centrifugation was then mixed with hot water and centrifuged again. The effluent from two centrifugations was combined and sent to refinery for processing[31]. According to Cambiella, a small amount of a coagulant salt $CaCl_2$ (0.01-0.5 M) can improve the water-oil separation process by centrifugation, with a high oil separation efficiency of 92-96%.

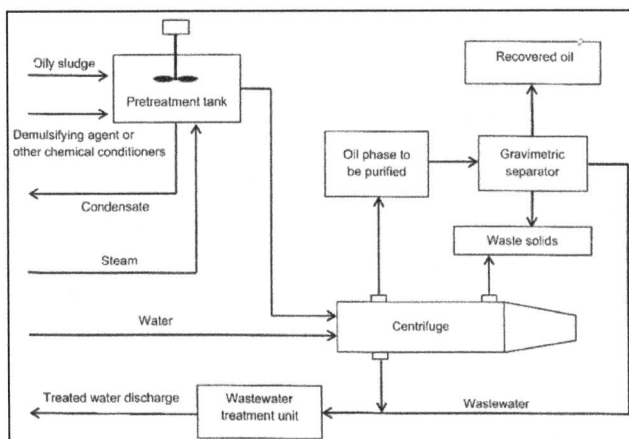

Fig: Schematic diagram of centrifugation method

Oily sludge is mixed with demulsifying agent or other chemical conditioners. The mixture is then treated by hot steam in a pre-treatment tank in order to reduce the viscosity. This less viscous petroleum is mixed with water for high speed centrifugation. The separated water after centrifugation containing high concentration of PHCs drained for further wastewater treatment. The separated oil containing water and solids is sent to a gravimetric separator for further separation to obtain the recovered oil. The separated water from the separator is sent to wastewater treatment. The sediments from centrifugation and separator are collected as solid residue for further treatment. Centrifugation is a relatively clean and mature technology for oily sludge treatment, and its oil separation from sludge is effective. Centrifugation equipment does not occupy large space. However in this method high energy consumption is required to produce high centrifugal force to separate oil from petroleum sludge. High equipment investment is responsible for the limited use of centrifugation method. The addition of demulsifying agents and tensioactive chemicals for pre-treatment increases the processing cost. Centrifugation process creates high noise.

Electrokinetic Method

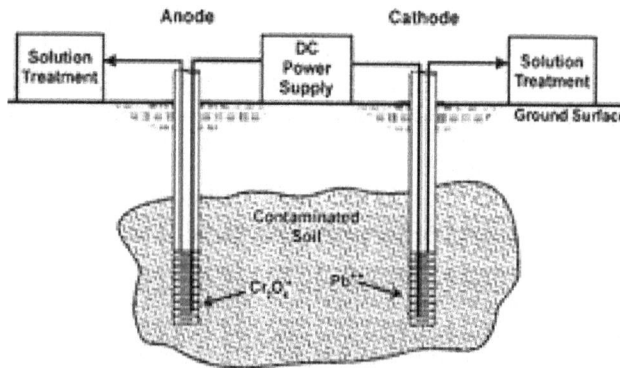

In this method, an electrode pair is used on two sides of a porous medium and a low direct current is passed through the medium causing the electro-osmosis of liquid phase, migration of ions and electrophoresis of charged particles in a colloidal system to the respective electrode. The separation of water, oil, and solids from oily sludge can be carried out by electrokinetic method and this separation is based on three mechanisms. Colloidal aggregates in oily sludge can be broken due to electric field and this leads to the movement of colloidal particles and solid particles of oily sludge towards the anode as a result of electrophoresis. Water and oil move towards the cathode as a result of electro-osmosis. The electro-coagulation of the separated solid phase occurs near the anode, this increases the concentration of solid phase and the sediments. The separated liquid phase (water and oil, without colloidal particles and fine solids) can form an unstable secondary oil-in-water emulsion, which could be gradually electro-coalesced near the cathode through charging and agglomeration of droplets; thus, forming two separated phases of water and oil.

Ultrasonic Irradiation

Ultrasonic waves generate compressions and rarefactions in the medium through, which they are passed. The compression cycle exerts a positive pressure on the medium by pushing molecules together. The rarefaction cycle exerts a negative pressure by pulling molecules from each other. Micro bubbles are produced in the medium and these will be grown due to negative pressure. These micro bubbles grow to unstable dimension and collapse violently generating shock waves, which results in high pressure and temperature immediately. This increases the temperature of the emulsion system and decreases its viscosity, increases the mass transfer of liquid phase, and thus leads to destabilization of W/O emulsion. Smaller droplets in emulsion move faster than the larger ones under the influence of ultrasonic irradiation. This can increase their collision frequency to form aggregates and coalescence of droplets, which then promotes the separation of water/ oil phases.

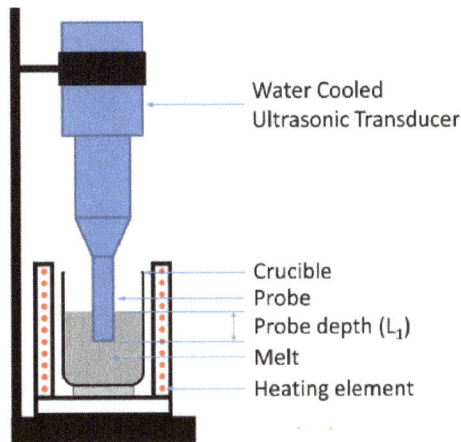

Froth Flotation Method

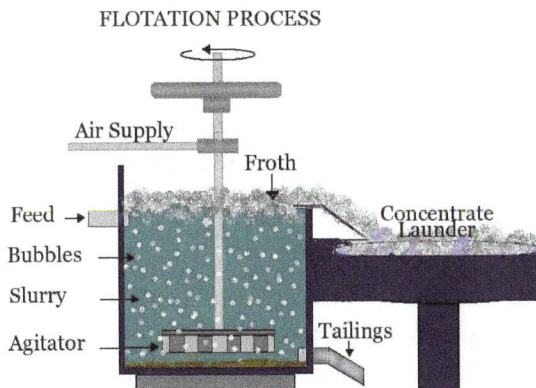

This method involves the capture of oil droplets or small solids by air bubbles in an aqueous slurry followed by their levitation and collection in a froth layer. Froth flotation process can be used for the treatment of oily wastewater from the refineries and oily sludge. In this method, water is mixed with oily sludge to form oily sludge slurry. Air is passed through the sludge slurry, which form air bubbles in the water sludge mixture. These air bubbles approach oil droplets in the slurry mixture. The water film between oil and air bubble becomes very thin and then it is ruptured causing spreading of oil in the air bubbles. The conglomerate of oil droplets with air bubbles can quickly rise to the top of water-oil mixture, and the accumulated oil can be skimmed off and collected for further purification.

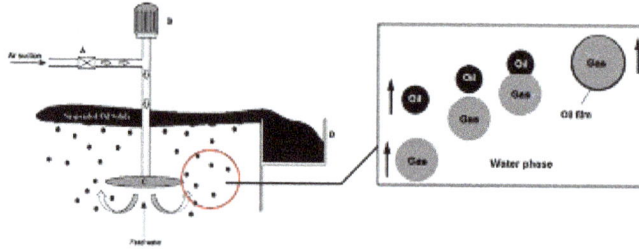

Fig: Schematic diagram of froth flotation process

Oily Sludge Disposal Method

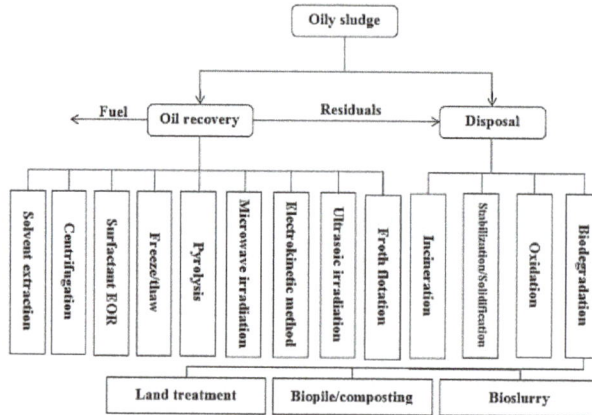

Oily sludge after the recovery of oil should be disposed of by a number of methods such as incineration, stabilization/solidification, oxidation and biodegradation.

Incineration

In this method, complete combustion of oily wastes is carried out in presence of excess air and auxiliary fuels. The commonly used incinerators are rotary kiln and fluidized bed incinerator. The combustion temperature range for rotary Kelvin is 980-1200° C and residence time is around 30 min. In fluidized bed incinerator, the combustion temperature can be in the range of 730-760° C and the residence time can be in the order of days43. Fluidized bed incinerator is more effective for low quality sludge due to its fuel flexibility, high mixing efficiency, high combustion efficiency and relatively low pollutant emissions.

Stabilization/Solidification

By this method, contaminants are immobilized by converting them into a less soluble or less toxic form (stabilization). The contaminants can be encapsulated by creating a durable matrix (solidification). Inorganic wastes are easily disposed of by this method. This method is less compatible with organic wastes.

Oxidation Method

Oxidation treatment is useful method to degrade a number of organic contaminants through chemical or other oxidation processes. Chemical oxidation is carried out by adding reactive chemicals into oily wastes, which oxidize organic compounds to carbon dioxide and water or transform them to other non-hazardous substances such as inorganic salts. The oxidation can be carried out by Fenton's reagent, hypochlorite, ozone, ultrasonic irradiation, permanganate and per sulphate, by generating a sufficient amount of radicals such as hydroxyl radicals (OH*), which can quickly react with most organic and many inorganic compounds.

Bioremediation

Refineries produce oil sludge as waste by the processing of crude oil. This sludge is poorly biodegradable. One of the major problems faced by oil refinery is the safe disposal of oily sludge in the environment. Many of the constituents of oil sludge are carcinogenic and potent immuno-toxicant. This technique uses living organisms (bacteria, fungi, some algae, and plants) to reduce or eliminate toxic pollutants. These organisms may be either naturally occurring or may be cultivated in the laboratory. They either eat up the contaminants (organic compounds) or assimilate within them all harmful compounds (heavy metals) from the surrounding; thereby, rendering the region con-taminant free. Bioremediation can be enhanced with the use of fertilizers, compost, bulking agents and some chemicals including oil dispersant. Bioremediation of waste oil in soil (land farming) or also land spreading had been carried out in different parts of the world. Dotson and Kincannon published early studies in USA. By mid 70's, other authors reported on oily sludge land farming including Fusey in USA. More studies on bioremediation by land farming were reported by a number of authors namely Grove in UK, Dibble and Bartha in South Africa in the late 70's.

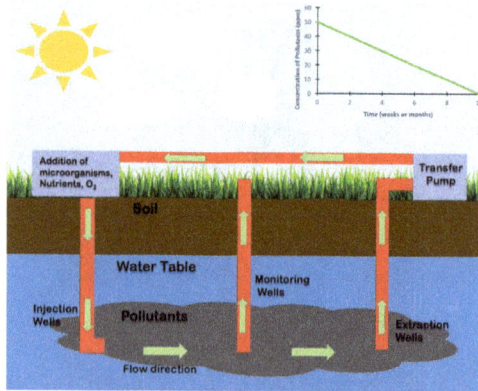

Land farming is cheaper and environmentally safe method. In land spreading, the sludge is evenly dispersed over a plot of land, where it can be degraded by native mi-crobial flora over a period of months or years. The sludge is blended into the soil with tilling equipment, and the rate of degradation in the land farming is increased with the addition of fertilizers. The primary mechanisms involved in the disappearance of hydrocarbons in land spreading and land farming are biodegradation, vaporization, oxidation, and to some extent, degradation by sunlight and leaching. When the sludge has been substantially degraded, the plot of land can be used again for further sludge treatment.

Rehabilitation of Storage Tanks

1. Rehabilitation of tanks for the storage of chemical and oil-based products, with recovery of products and reduction of waste. Dregs form in tanks of liquid chemical and oil-based products.

    Companies use innovative technologies for the rehabilitation of tanks, prioritising operational and environmental safety, recovering the product from the dregs and reducing the weight and volume of the residues to dispose of:

    ○ JET MIXER System: a highly powerful mixer system which homogenises the liquid dregs leaving the heavy, non-reusable material at the bottom. The system requires many operating cycles, which can include cleaning of the dregs (skimming), and allows safe entry to qualified personnel for removal of the elements that cannot be pumped.

    ○ BSW System: a system for treating the dregs that separates the water and solids in the crude oil dregs.

    ○ This technology allows the recovery and re-use of over 95% of the product, significantly reducing the volume, weight and cost of the material to dispose of.

    ○ SPOT System: a system for monitoring and quantifying the dregs that gives the customer a clear idea of the content and quantity of the dregs deposited in the tanks.

2. Rehabilitation and cleaning of equipment, machinery and pipelines.

3. Decommissioning of industrial equipment and salvage of metals.

4. Rehabilitation of contaminated sites using advanced technology.

Treatment of Semi-processed Oil Refinery Products

Oily sludge resulting from oil refinery processes (API tanks, tank sludge, biological sludge from the treatment of oily sewage waste) are difficult products to manage due to their low fluidity and their high hydrocarbon content.

This kind of sludge, if not treated, due to its high oil content, can only be delivered as it is to treatment plants or to incineration. This solution has important costs and technical implications, due to the need to move a mass whose consistency is between pump able and shovel able, which also is often be classified as hazardous for transport and, for handling. Moreover, in this way it will not be possible for the Customer to recover the hydrocarbon fraction from the sludge.

To meet this challenge, Ecomed has developed a system for the management of semi-processed products aimed at maximizing oil recovery while at the same time reducing the residual fraction of the treatment, to be sent for disposal off site.

The system used is divided into two phases:

- The treatment of semi-processed oil refinery products, which results in:

  ○ liquid flow (composed of water and hydrocarbons), returned to the Customer for reintroduction into the refining cycle.

  ○ a solid flow, which is classed as waste and sent for further stabilisation/solidification treatment.

- The treatment of solid waste, resulting in a semi-solid mass, suitable for safe transport for offsite disposal.

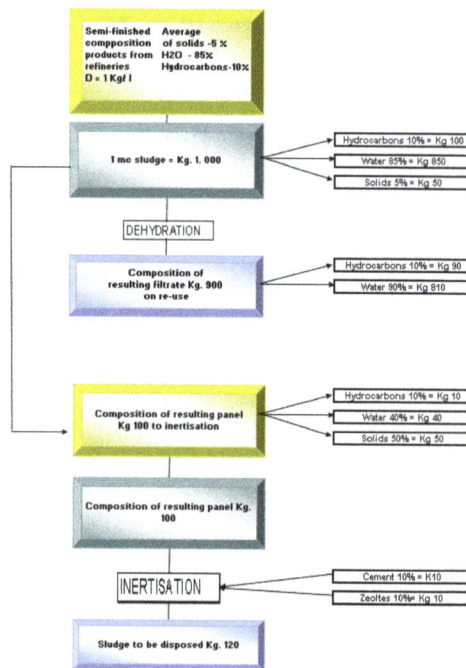

The process applied consists of:

- dehydration treatment of semi-processed oil refinery products;

- stabilisation/solidification treatment of solids derived from the dehydration of semi-processed products.

The dehydration treatment includes:

- accumulating the sludge in a homogenization tank, where it is mixed and heated;

- supplying the sludge to the dehydration system through a single screw pump with variable capacity;

- the dosage on the line of appropriate polyelectrolyte selected for the typology of the sludge supplied and of the result required;

- phase separation through centrifugation of the conditioned sludge.

According to requirements and on the Customer's request, the process can be optimised to obtain a drier panel but with higher oil content, rather than a panel with a slightly higher humidity after recovery of a higher percentage of oil.

The stabilisation/solidification treatment can be carried out on the same line or on a specialised system. To be able to discharge the waste, stabilisation is carried out by mixing in water binders (lime or concrete) with the waste and additives which adsorb the oil residues (zeolites, urasites) and, if the panel needs to be sent for thermal treatment, organic amendants will be added (sawdust, coconut fibres). However, the system is sufficiently flexible to adapt the reagents used to various requirements without the need to modify the installation.

Remediation of Hydrocarbon Contaminated Soils

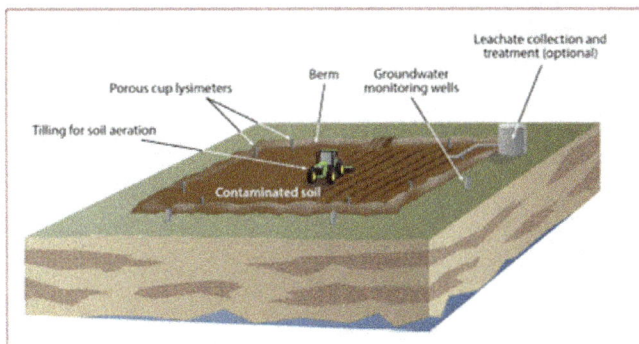

The goal of remediation is to remove, or to make harmless, substances contaminating the soil or groundwater. The remediation processes can be applied directly to the site of contamination, in situ, or after removing the contaminated soil, ex situ. Among the latter, there are the treatments on site, when working on the excavation site or off site, when you need to transport in plants located elsewhere. It is possible a classification of the different processes in accordance with the mechanism for cleaning up: non-organic (chemical, physical or thermal) or biological. The choice of different types of treatment is linked to several factors related to the nature of the pollutant, the polluted site, the type of technology (basically to its efficiency and cost). Briefly, the main techniques of both non-biological and biological remediation will be described below.

Non-biological Remediation

Soil Washing

This technique, generally carried out on site, consists in the soil decontamination by washing with water and possible addition of other substances (chelating agents, surfactants, acids or bases), according to the needs, in order to improve the 'solid-liquid extraction in contact with the ground. The latter undergoes an initial pre-treatment to remove coarse material and then switch to the washing stage where there is the process of extraction-solubilisation of the pollutant, which passes into the aqueous solution. Subsequently it has a solid-liquid separation that provides the clean solid and starts the water to the treatment thanks to which the pollutant is separated, concentrated, and recirculates the process water for the next wash. The capacity of reclamation is quite high (about 25 t / h). The duration of treatment Soil Washing is usually short, from one to three months and increases as the percentage of clay and silt content in the soil.

Development and applicability: currently, it is developed on a large scale and found more applications in Europe than in America especially in the removal of heavy metals. It can be applied, however, to a wide range of pollutants, including hydrocarbons and pesticides.

Critical issues: May be limited by silty and clayey soils that make it more difficult to solid-liquid separation. Physical separation is generally not effective for treating the chemically adsorbed metals.

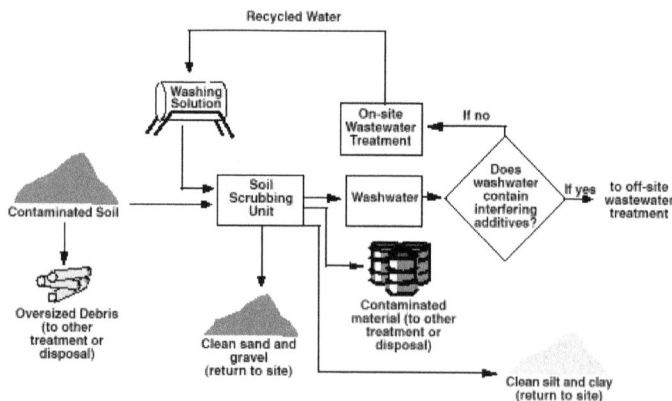

Figure: Soil Washing

Solvent Extraction

Technique operating on site using a solvent to improve the efficiency of extraction, in a process very similar to that previously described. Since traces of solvent may remain in the ground at the end of the treatment, a criterion for its choice concerns the degree of toxicity.

Development and applicability: the solvent extraction has proven effective in removing a wide range of organic pollutants, from hydrocarbons to organ chlorine pesticides, VOCs and petroleum wastes. The plants in full scale come to treat 20t/h of soil.

Critical issues: It is not applicable for the removal of inorganic pollutants and some processes are limited by the solid matrix moisture content and fine particles. The presence of detergents and emulsifiers can unfavourably influence the extraction performance.

Soil Vapor Extraction (SVE)

This technique, also called Soil Vacuum Extraction, is applied in situ and used in the reclamation of the unsaturated zone of the soil, the area in which the pores of the soil contain air or water at a pressure lower than the atmospheric one (by capillarity). Using a system of wells, vacuum is applied so as to induce a controlled flow of air from outside which brings with it the volatile compounds and some semi volatiles. This system comprises a gas treatment extracts made from activated carbon filters, systems of incineration or cold traps; the treated gas is released into the atmosphere or re-injected into the ground.

Development and applicability: This method finds application mainly in soils at medium depth and permeability to avoid the short-circuiting of the steam flow or a difficulty in its circulation. The pollutants to be removed must have a vapor pressure greater than 1 mm Hg at 20°C. Both SVE and air sparging are used to clean up several acres of contaminated soil and groundwater at the Vienna PCE Superfund site in West Virginia.

Critical issues: Soils with moisture content above 50%, tend to adsorb the pollutant compounds, compromising the effectiveness of the technology.

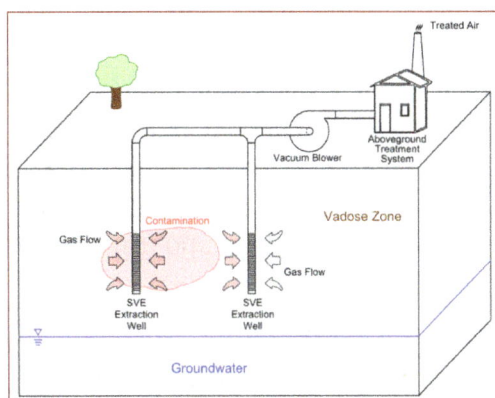

Figure: Soil Vapor Extraction (SVE) System for Vadose Zone Remediation Air sparging

Air Sparging

Technology operating in situ in which air is bubbled through a contaminated aquifer. The air bubbles cause stripping of volatile organic compounds present in the saturated zone, the part of the subsurface in which the pores of the soil are filled with water at pressure equal to or greater than the atmospheric one. In general, the exit gas from the underground are conveyed by means of a suction system, which often coincides with a SVE inserted into the unsaturated zone.

Development and applicability: it finds ideal application in homogeneous soils with high permeability and unconfined aquifers, polluted by volatile compounds, halogenated and hydrocarbons.

Critical issues: if it is not used in conjunction with SVE, an unwanted migration of pollutants outside the contaminated area may occur. Special attention should be given in the event of large doses of pollutant supernatants (eg. Hydrocarbons in suspension), to prevent the push and the bubbling cause aerosols in the surrounding areas.

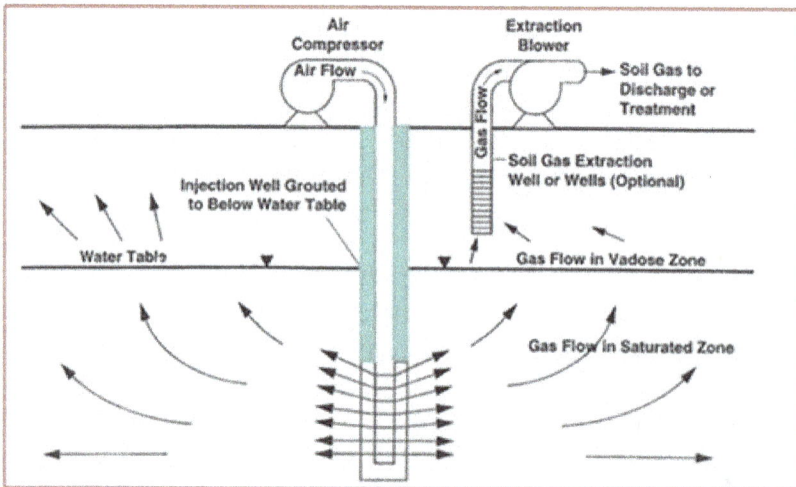

Figure: Combined Air Sparging and Soil Vapor Extraction system

Dual Phase Extraction

This technology allows the in situ simultaneous removal of the contaminants present in the unsaturated zone and the saturated zone of the soil (In case the contamination concerns both stages) by the means of a vacuum pump. In this way extends the applicability of the SVE to the saturated zone of the soil. Downstream of the vacuum pump it is necessary to separate the liquid from the vapour phase and proceed to the train of treatments for the different phases.

Development and applicability: suitable for this technology are the low permeability soils, usually clay, in which the cone of depression extends in depth, going to increase the thickness of the unsaturated zone. An important factor to consider concerns the hydrogeology of the site, crucial to understand the degree of applicability and effectiveness of this treatment.

Critical issues: The technique is not recommended in case of deep contamination of the aquifer and in the case where the contamination is very extensive.

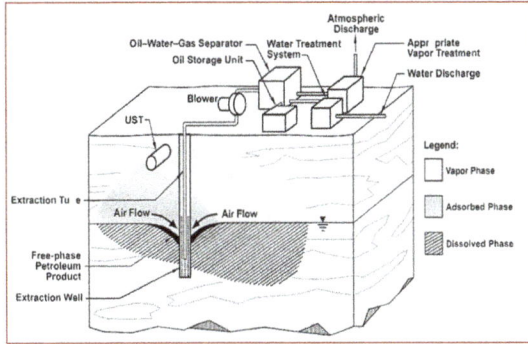

Figure: Typical Dual Phase Extraction Scheme

Solidification / Stabilization

In the solidification processes, the pollutants are physically linked, or trapped in a solid matrix, while in the stabilization, chemical reactions transforming the pollutant in a less mobile species, are favoured. An example is given by the cement that immobilizes many metal contaminants by forming insoluble hydroxides, carbonates and silicates (stabilization) as well as providing an encapsulating matrix for the leaching attenuation (solidification).

Development and applicability: this technique is used mainly for the treatment of inorganic pollutants, including radionuclides, while the presence of organic material may constitute an obstacle for the success of the neutralizing process.

Critical issues: increase in volume of the final product (up to twice the volume to be treated) and long-term stability of the material inertization.

Thermal Desorption

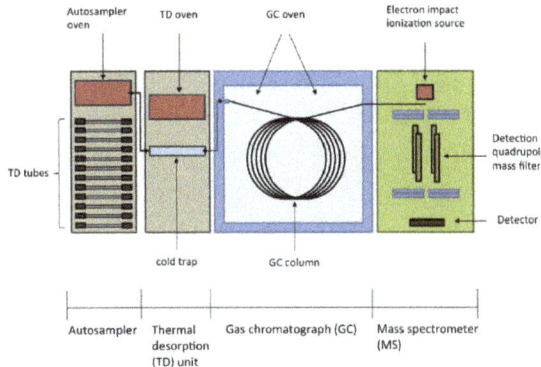

This ex-situ treatment consists in the desorption of volatile pollutants through the supply of heat from outside. The material polluted, is sent to a rotary kiln or to a heated auger system, where, by increasing the temperature the formation of gases and vapours of polluting compounds is guaranteed. The contaminants destruction is realised using a secondary treatment units.

Development and applicability: there are two processes in response to operating temperature, low (90-320°C) and high (320-560°C). In the first case, generally suitable for non-halogenated hydrocarbons, no thermal oxidation occurs and the physical characteristics of the soil remain unchanged. In the second case, instead, semi volatile organic compounds, volatile metals and polycyclic aromatic compounds are removed by operating, often in combination with the incineration processes, solidification/stabilization and dechlorination.

Critical issues: the economy of the process is affected by the moisture content of the material to be reclaimed.

Incineration

Technology that works off site used for the final disposal of contaminated materials resulting from the treatment of soil washing, solvent extraction and thermal desorption. The contaminated material is fed into a burner where takes place the volatilization and oxidation of organic compounds at temperatures between 870°C and 1200°C, in the presence of oxygen. Often it is necessary to supply the burner with an auxiliary fuel, both to trigger and to maintain the combustion.

Development and applicability: used especially if contamination concerns explosives, chlorinated hydrocarbons, polychlorinated biphenyls and dioxins. In the presence of heavy metals, it is necessary to inert ashes.

Critical issues: at the exit of the burner, gas treatment, particulate abatement and neutralization of acids (HCl, NOx and SOx) systems are needed.

Biologic Remediation

Bioventing

Technology operating in situ, for the soils treatment in the unsaturated zone. It stimulates the degrading action of microorganisms already present in the soil (native microbial flora), providing oxygen and, where necessary, mineral nutrients into the ground by percolation or by direct input with specific spargers. Oxygen is normally provided through direct input or air suction through spears stuck in the ground.

Development and applicability: it is useful in the remediation of hydrocarbons contaminated soils and is adaptable to soils with high permeability. Process often coupled with SVE: first making an SVE with the removal of the more volatile hydrocarbons, then performs a bioventing, simply by reducing the air flow and injecting nutrients for degrading residual non-volatile hydrocarbon components.

Critical issues: avoid the application when soils are: too heterogeneous, so contaminated by pollutants to create a saturated zone and in areas close to aquifers.

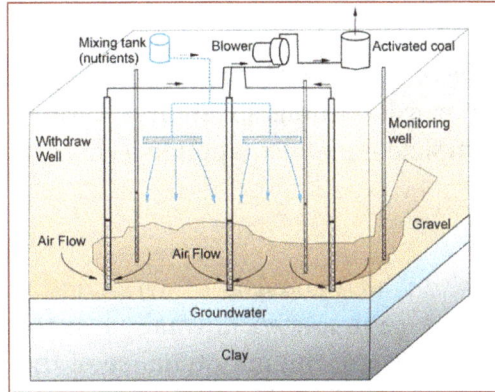

Figure: Typical Bioventing system

Biosparging

Technology similar to bioventing, which operates in situ for the treatment of saturated soils and groundwater. The acceleration of the native microbial flora degradation is done through direct air and appropriate nutrients entering the contaminated area.

Development and applicability: it is normally used to degrade the contaminants that are dissolved in the groundwater, adsorbed on soil particles below the groundwater level or in the capillary fringe area. Effective is the application in petroleum products reduction, usually realised for underground storage tank sites.

Critical issues: the presence of an hydrocarbons concentration such as to decrease the permeability or to be toxic, can make ineffective the action of biosparging.

Natural Attenuation

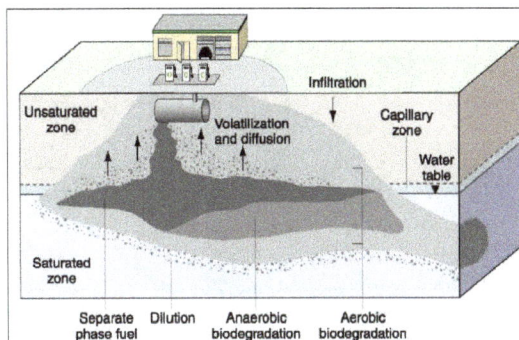

In this way is performed an intrinsic bioremediation, exploiting the nature ability to restore a polluted environment. Set up a site for a natural attenuation essentially means: run a targeted monitoring to know the precise boundary between the contaminated area and the clean zone, a campaign of analysis for the measurement of some basic parameters (temperature, pH, redox potential, concentration nitrate, nitrite and ammonia, phosphates) and enumeration of bacterial populations specific for the different types of biodegradation.

Critical issues: the presence of non-biodegradable pollutants, existence of contamination phenomena able to convey hazardous substances towards targets of environmental interest and need to complete the remediation in a short time.

Land Farming

Technique operating on site which consists of arranging the contaminated material on a non-permeable surface in a layer normally less than one meter, ensuring, during the decontamination period, the maintenance of the best conditions for the microbial populations development. It is essential to ensure, from the beginning of the treatment, a correct balance of the main nutritional components of the system: carbon, nitrogen and phosphorus, in relations respectively 100: 5: 1 in addition to the content of water content (60-70% of the saturation value), and the soil pH, which must be neutral. Furthermore it is necessary to facilitate the air entry for the correct oxygen supply to the bacterial populations, generally by mixing the soil to be treated or by entering bulking agents (wood chips, expanded silicon, etc.). This process requires an extended time frame, possibly up to 24 months, depending on a number of factors, including the nature of the contamination, the concentrations of contaminants, types of soil and volume of soil to be remediated. Land farming to remove volatile constituents from soils through evaporation, without biological degradation, is not acceptable, unless are realised the volatile constituents capture and treatments.

Critical issues: the presence of volatile pollutants that threaten the operators health, unavailability of adequate land area.

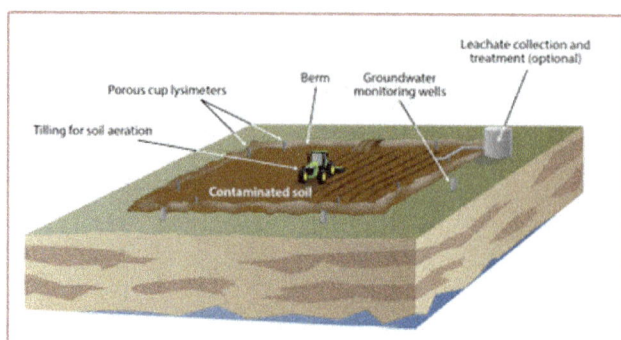

Figure: Typical Land farming operation

Biopiles

Technique very similar to land farming, with the main difference residing in the method of oxygen transfer. In the preparation of the biopile soil layers are superimposed with interspersed perforated tubes, used to distribute, air and solutions containing the necessary nutrients. In the presence of volatile pollutants, the biopile can be covered with waterproof sheets with appropriate openings to let out the steam to be sent to treatment. The presence of the sheets also facilitates the monitoring of the parameters indicated in the land farming section.

Development and applicability: The application of biopiles has been guaranteed by numerous studies that have shown good removal efficiencies especially for total petroleum hydrocarbon (THP).

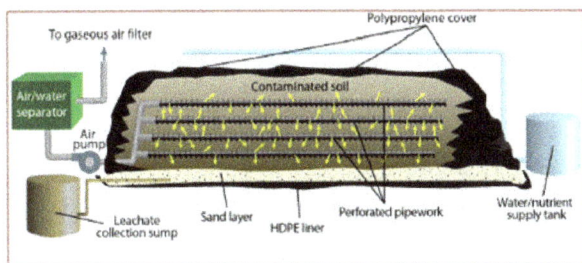

Figure: Typical Biopiles Scheme

Bioslurry

Technique consisting in the remediation of soil within fermentation reactors. Inside the reactors can be controlled effectively operating parameters or use non-native bacterial populations (bioaugmentation). Theoretically it would be possible to use genetically modified bacteria, practice today banned in open field.

Development and applicability: to date, this method has been applied only to remove substances not readily degradable.

Critical issues: high costs and reduced volumes of treated soils limit the operational capabilities of bioslurry.

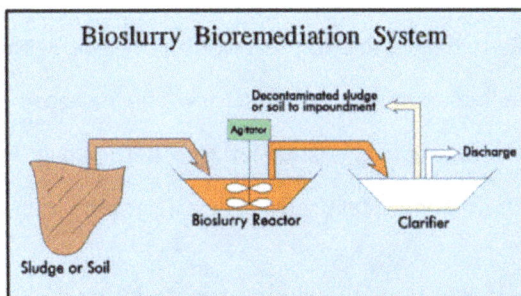

Figure: Bioslurry system

Microbiological Barriers

A system for the in situ treatment of groundwater. Inside the aquifer is placed transversally a barrier consisting of soil or suitable solid support colonized by microorganisms. Therefore the barrier is "passive", and biodegradation occurs by contact between the water that runs through it and microorganisms adhering to it. It is also necessary to construct a series of wells for the air and nutrients intake for the present microorganisms.

Development: the microbiological barriers, have had a good spread in Italy with some application examples.

Emerging Technologies

In recent years, research in the field of bioremediation is evolving to try to increase the contamination cases resolved by biological remediation. In this regard two research lines are being developed to exploit synergistic actions respectively between microorganisms and plants (bioremediation/phytoremediation) and between bacteria and fungi. Not to be overlooked are the research of genetics to modify microorganisms and make them capable of degrading substance considered recalcitrant and the research to make possible actions in situ even in situations where it is very difficult to convey sufficient amount of oxygen. Have particular importance the cases of hydrogen or magnesium peroxides use.

In the non-bio field, permeable reactive barriers (continuous and non), characterized by high conductive reactive media capturing the contaminants by deviating the natural flows, are producing invigorated research results. Another interesting field of research can be individuated in the use of hydraulic fracture technology for soil and groundwater remediation. FRx promotes the Hydraulic fractures as an optimal in situ treatment for the removal of materials from extraction wells. They utilizes US EPA practices and individuates four key factors distinguishing the uses of hydraulic fractures for soil and groundwater remediation from those ideated for the oil and gas production: volume, depth pressure and chemical additives.

References

- Petroleum-sludge-its-treatment-and-disposal--a-review: tsijournals.com, Retrieved 16 May 2018

- Storage-tanks-rehabilitation: effegigroup.it, Retrieved 30 May 2018

- Refinery-waste-treatment-services: ecomed-intl.com, Retrieved 25 June 2018

- Remediation-of-hydrocarbon-contaminated-soils: oil-gasportal.com, Retrieved 15 July 2018

Chapter 5

Textile Waste Treatment

The textile industry is concerned with the production, design and distribution of garments, cloth and yarn. The manufacturing of textiles results in large amounts of hazardous waste. The topics elaborated in this chapter such as silk waste, textile recycling, textile waste and textile wastewater treatment will provide a detailed understanding of waste, generated by the textile industry and its treatment.

Textile Waste

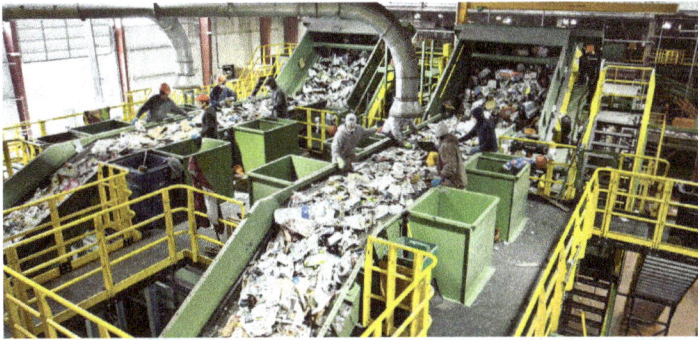

In recent years, textile production and consumption have risen drastically due to global population growth and improvements in living standards. Over-production in the textile industry is partly driven by the idea behind the fashion industry that consumers need a new clothing collection for each season. This increases the rate of replacement of the products and the rate of textile and waste generation. For example, during the last 10 years in Sweden, textile consumption has increased by 40%, and the annual clothing and home textile consumption is now 15 kg per capita. The consumption of textiles in other Nordic countries ranges between 13.5 kg per capita in Finland and 22 kg per capita in Norway.

Clothing and textiles are produced from fibres which are either natural (e.g. silk, wool), natural cellulosic (e.g. cotton, linen), manufactured cellulosic (e.g. viscose, rayon) or synthetic (oil-based e.g. polyester, acrylic and nylon). Figure, shows the global consumption of fibres between 2000 and 2012. It can be seen that the global consumption of fibres reached a short-term maximum in 2007. Textile fibre consumption decreased by 4.3% to 66.1 Mt in 2008 due to the global financial recession (FAO/ICAC 2013). The demand for all textile fibres in 2012 was 74.8 Mt, which is an increase of 13% over 2008.

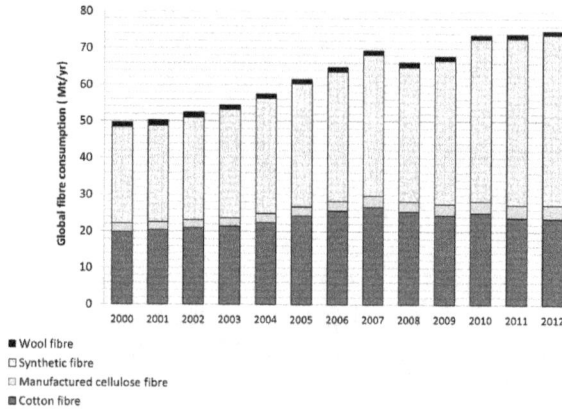

Figure: Annual global fibre consumption (2000-2012).

The global consumption of cotton followed the same trend as the total global fibre consumption, which hit the highest point in 2007. Total demand for textiles fell in the following year but has since recovered and exceeded this short-term peak. However cotton demand has not recovered since then. The share of cotton fibers in global fibre consumption was falling from 38% in 2007 to 31% in 2012. Over the whole 2000-2012 period, cotton consumption increased by only 18%. The reasons for this decrease are the instability in cotton prices due to rising labour costs in China and cotton's competition with food production for irrigable land and water. Based on a recently published report from OECD, it is predicted that world cotton consumption is expected to recover relatively slowly. However, total cotton consumption is predicted to reach 30.8 Mt in 2023 when it will exceed the 2007 peak.

The consumption of synthetic fibres had increased by 77% between 2000 and 2012. The growing share of synthetic fibres in global fibre consumption has resulted in a rising demand for petroleum-based chemicals. For this reason, attempts at developing new oil extraction technologies have become significant for the textile industry. One innovative method is hydraulic fracturing of rock ('fracking') which keeps the price for petroleum-based chemicals and synthetic fibres low.

The consumption of manufactured cellulosic fibres increased by 63% between 2000 and 2012. Although the current rate of demand for manufactured cellulosic fibres is not fast enough to offset the production rate for cotton, it indicates the keen demand for cellulosic fibres with similar properties to cotton.

In this context, different approaches have been suggested for increasing cellulose-based fibre production:

1. Different technological alternatives for cotton production: Genetically Modified (GM) cotton could compensate for reduced Chinese cotton supplies to some extent. This entails using genetic engineering techniques in which the genetic coding for a common soil bacterium, Bacillus thuringiensis (BT), is inserted

into cotton to promote the production of a natural insecticide in cotton tissues. This approach has been implemented in China, and after commercialisation, the plantation area for BT cotton in China increased from 16,700 hectares to 3.8 million hectares in 2007, which is 69% of the Chinese cotton farming lands. Studies show that by implementing this method, the yield of cotton plantations can increase up to 24%.

2. The development of new forest-based fibres: one example is Cellunova which is a regenerated cellulosic fibre from wood pulp. The newly developed fibers can be blended with cotton fibers since it they have cotton-like quality.

3. Recycle and reuse: One solution is to turn the discarded cellulosic textiles into a new resource by developing technologies for the mechanical reuse of textiles or for chemical recycling of fibres.

Current Textile Waste Management

When consumers decide to give up their garments, they have a number of choices: discard, sell, or donate to use textile collectors such as charity organizations, municipalities, retail collectors or professional collectors. In Europe, discarded textiles are either incinerated or landfilled together with municipal solid waste. The donated or sold textiles are sorted and sent afterwards to reuse or recycling plants depending on their quality. Most of the textiles which retain high enough quality for reuse are sent to East European or African countries (Palm et al. 2014), and the remaining flow is sent to recycling plants. Since only a few different methods for textile recycling exist today, the majority of the flow is down cycled into wipes, rags or is used as insulation in different industries. The remainder of the collected used textiles is either landfilled or incinerated. In some cases, clothing which is no longer in use is accumulated in closets or exchanged informally between friends or family members.

Presently, only 25% of the textile waste generated in the European Union is collected by charity organizations or industry enterprises with the purpose of reusing or recycling. The rest is sent to landfill or municipal waste incinerators.

Textile waste management in the UK

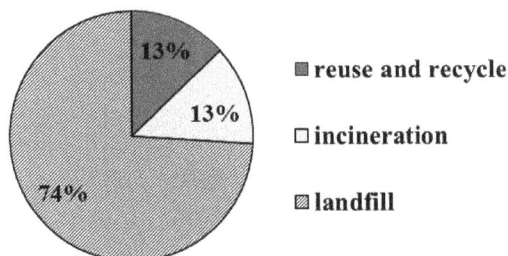

Figure: Textile waste management in the UK

Currently, a lack of practical technologies for recycling various types of fibres limits the potential for applying recycling techniques. However, a few small-scale recycling schemes have been implemented, with the aim of producing recycled fibres.

There are differences between textile waste management situations in many countries within the European Union. For example, as shown in figure 74% of the textile waste flow in the UK ends up in landfills, while 13% is sent to incineration. The remaining 13% is either eused or downcycled into wipes, upholstery fillings or insulation materials. In Sweden, approximately 20% of textile waste is collected by charity organisations, 50% is sent to incineration plants in mixed waste streams and the remaining 30% is either stored in household closets or discarded at recycling centres.

The statistical data are not fully comparable between the UK and Sweden since the definition of textile waste is different in each reference. However it can be noticed that landfilling has been removed as a textile waste management strategy in Sweden and the majority of the waste is sent to incineration for energy recovery.

The updated target of the European Union´s landfill directive entails forcing the re-duction of biodegradable municipal waste landfilled to 35% of the 1995 level by 2020 for the UK. Therefore, the UK faces challenges regarding the collection and recovery of biodegradable waste including textiles.

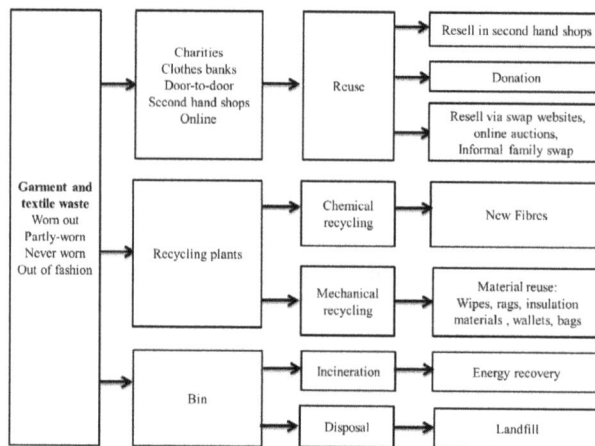

Figure: Strategies that can potentially be applied for textile waste management

Figure shows possible post-consumer routes for textile waste management.

Energy Recovery

Incineration with energy recovery is the dominant textile waste treatment technology in some countries, for instance Sweden. Collected textile waste from bins and sacks is sent to incineration together with other collected municipal waste. The recovered heat and power can potentially replace other sources of energy.

Reuse of Textiles

A share of the textiles and garments collected by charity organisations or collectors is transferred to second hand shops with the aim of sending used clothes into the market. Large amounts of clothing of adequate quality are shipped abroad for selling to other traders in Eastern Europe or Africa. 26,000 tonnes of collected used clothes and shoes in Sweden were donated to Africa and Eastern Europe in 2008.

Another example of the reuse of textiles are the new concepts of reselling or swapping second hand clothes through websites and online auctions that have emerged recently with the aim of extending the life span of garments.

Recycling Textiles

Another option for potentially saving resources in waste management is recycling. There are several different technologies available or under development for recycling textile waste.

Mechanical Recycling of Textile Waste

Different mechanical techniques exist for recycling textile waste. The applicability of each technology depends on the quality of the textile waste.

The most common mechanical recycling method is to cut and shred the fabric into small pieces which can be used as filling in mattresses or upholstery, as insulation or as carpet underlay. The SOEX group has facilities in 10 different countries that apply mechanical recycling techniques to convert more than 15000 tonnes of used clothing per year to insulation materials for construction and automobiles.

Another mechanical recycling method is material reuse without shredding. In this process, pieces of textiles with sufficiently high quality are separated and turned into different types of products. Small-scale upcycling enterprises, such as Worn Again and Loopt Works, produce T-shirts, wallets and textile carrier bags. This process is also known as upcycling or remanufacturing.

Another kind of mechanical recycling is the production of fibres and yarns from discarded textiles. The textile waste is first cut into small pieces, then passed through a rotating drum and turned into fibres. The physical quality of the fibres produced using this method is low due to the mixed colour of the fibres and the different fibre lengths. Therefore, the obtained fibre can only be used in upholstery filling, carpet underlay, sound and heat insulation materials, disposable diapers, napkins and tampons. One way to improve the quality of this product is to mix these fibres with virgin fibres and blend them into yarns. Since the properties of such yarns are dependent on textile quality, they are mostly used in producing woven filtration systems or geotextiles.

Chemical Recycling of Textile Waste

In principle, Chemical recycling methods can be applied to synthetic fibres (polyester, nylon or polypropylene) or blends of natural and synthetic fibres. During chemical recycling processes, the fibres are chemically separated and degraded to the molecular level. The synthetic feedstock is then repolymerized to new fibres.

In 2000, a Japanese company, Teijin Fibre Ltd., developed a closed-loop polyester recycling technique process in collaboration with Patagonia Inc. Initially, the main goal of the process was 100% polyester uniforms. In this process, polyester fabrics from other types of fabrics such as polyester/cotton, acrylic, wool, polyurethane or leather are completely separated. The collected material is then cut into small pieces and broken down into small granules. By applying a chemical reaction, the granules are decomposed to dimethyl terephthalate (DMT), an intermediate chemical for the production of polyethylene terephthalate (PET). The actual chemical reaction applied by Teijin has not been published, but a fair assumption is that the reaction is the depolymerization of polyethylene terephthalate in methanol, see Figure. The obtained DMT is repolymerized to polyester granules and turned into polyester yarns through melt spinning.

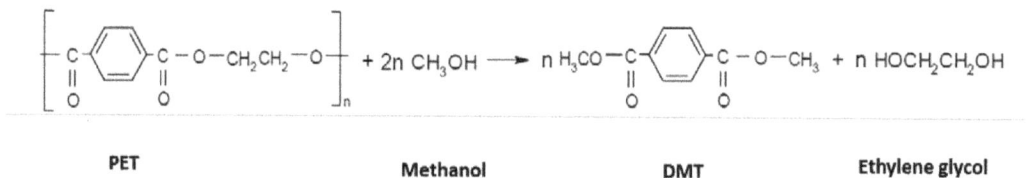

$$\left[\underset{O}{\overset{\parallel}{C}} \text{—} \bigcirc \text{—} \underset{O}{\overset{\parallel}{C}} \text{—O—} CH_2CH_2 \text{—O} \right]_n + 2n\ CH_3OH \longrightarrow n\ H_3CO \text{—} \underset{O}{\overset{\parallel}{C}} \text{—} \bigcirc \text{—} \underset{O}{\overset{\parallel}{C}} \text{—O—} CH_3 + n\ HOCH_2CH_2OH$$

PET   Methanol   DMT   Ethylene glycol

Figure: Chemical equation of methanolysis of PET

Emerging Technologies for Textile Waste Management

There are a number of emerging technologies for the chemical recycling of textile waste. One example is a process developed by the Swedish company Re:newcell, which regenerates viscose fibres from discarded cellulosic textiles. During this process, discarded textiles are mechanically cut into small pieces and the cellulosic share of the textiles is chemically separated, dissolved into an alkali solution, and, finally, filtered to produce regenerated viscose. Small-scale tests on different types of blended textiles have been conducted. However, little information is available on the process as it is still under development.

There are currently ideas for using the Lyocell process for recycling cellulosic and synthetic fibres from discarded textiles. In the Lyocell process, cellulose in wood pulp is dissolved with N-methylmorpholine-N-oxide (NMMO). The solution is filtered and cellulose fibres are extracted. Cellulose fibres are then washed and spun into yarns. The NMMO is recovered and sent back to the process for reuse.

It has been suggested that separating and recycling cellulosic and synthetic fibres by using NMMO can be developed further. In the first step, blended textiles are cut into pieces and mixed with NMMO. The cellulosic part dissolves into NMMO and is pumped through filters to separate the solution from polyester residue, and the remaining polyester can be recycled into polyester fibres.

Environmental Impacts of Textile Waste Treatment

The reuse and recycling of discarded textiles has several potential environmental benefits. For example, reusing textile and clothing products results in energy savings, since the amount of energy required for collecting, sorting and reselling second hand garments is 10 to 20 times less than for the production of the same products from virgin materials (Fletcher 2008b). Since the benefits vary so much, the quantitative assessment of alternatives is warranted. In order to quantify the environmental impacts of alternative textile waste management systems, there are a number of factors which must be assessed:

- The demands for energy and material during the mechanical, chemical or biological processes in the textile recycling value chain; and

- The emissions into air, water and soil that occur while applying textile recycling processes. Thus, for evaluating the benefits of the recycling schemes, one must calculate whether the consumption of energy and resources and the emission of pollutants by the recycling process are compensated for by avoiding the manufacturing of products from virgin materials.

The first part of this thesis assesses the environmental benefits of three textile recycling techniques in order to investigate the options for improving textile waste management. For this purpose, an environmental life cycle assessment (LCA) was carried out to quantify the energy usage and global warming potential of different textile recycling options. The studied technologies are: remanufacturing for material reuse, chemical separation of cellulose from polyester using NMMO, chemical recycling of polyester, and incineration with energy recovery.

Social Issues Related to Textile Waste Treatment

Existing Social Issues in the Textile Industry

Consumer expectations on low price products and companies' competition for market share result in a number of social violations. Some of the significant issues in the textile and clothing value chain include low worker wages, gender discrimination, excessive working hours, temporary working contracts or child labour and local residents who are subjected to health risks.

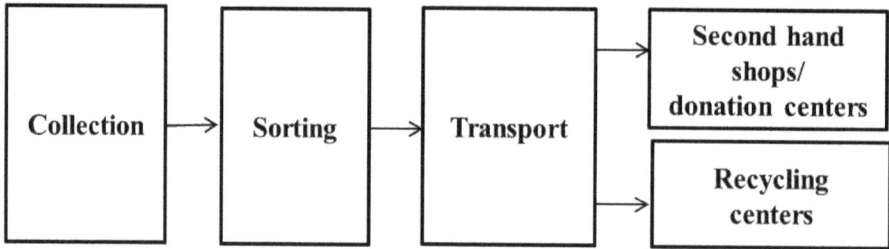

```
┌──────────────┐   ┌──────────────┐   ┌──────────────┐        ┌──────────────────┐
│              │   │              │   │              │   →    │   Second hand    │
│              │   │              │   │              │        │      shops/      │
│  Collection  │ → │   Sorting    │ → │  Transport   │        │ donation centers │
│              │   │              │   │              │        └──────────────────┘
│              │   │              │   │              │        ┌──────────────────┐
│              │   │              │   │              │   →    │    Recycling     │
└──────────────┘   └──────────────┘   └──────────────┘        │     centers      │
                                                              └──────────────────┘
```

Figure: Phases in the life cycle of textile waste reuse/ recycling

Figure presents the life cycle of textile and clothing waste reuse and recycling. Both reuse and recycling approaches create employment in each phase indicated by a box.

A share of the collected textile and clothing waste is shipped abroad to be donated or sold as second hand clothes. The second hand clothing industry provides a number of jobs in the receiving countries during transportation, distribution, cleaning, repairing and restyling. For instance, approximately 150,000 Ghanaians work in second hand clothing sector in Africa.

The overall impact of the second hand clothing industry on employment in developing countries is not clear, since it is likely that it plays a role in the decline in domestic textile and clothing production. In West Africa, up to 80,000 jobs were lost in the local textile and clothing industry between 1990 to 2004 (Baden and Barber 2005). However, the pressure on the local textile and clothing industry is not only caused by the second hand clothing sector.

Another factor is the import of cheap clothing from China.

An additional concern is the poor regulation of second hand clothing imports in developing countries. Corruption is known to occur when new garments are imported under the label of second hand clothes to avoid the payment of the appropriate, higher import tariffs for new clothing.

Finally, working conditions in the textile reuse and recycling sectors are other important social concerns. These sectors face challenges in dealing with working conditions in developing countries. In Senegal, 24,180 people have full time employment in the second hand clothing sector, of which 60% of are male. Gender discrimination, adequate wages, reasonable working hours, health and safety issues must be assessed along the entire value chain of the industry.

The complexity of the textile and clothing industry has made it difficult to assess the social issues along the supply chain. Emerging ecolabels´ claims for textile products cover a variety of different indicators. Some examples of indicators are natural resource management (e.g. Global Organic Textile Standard, Textile Exchange and Soil Association), restriction on chemical usage (e.g. Blue sign, Svanen and EU Ecolabel), improved working conditions in the textile and fashion industry (Fair Wear Foundation) and health issues for consumers (e.g. OEKO-Tex). Each label does not cover all indicators, but the major ecological issues are considered by most of the labels.

Silk Waste

Silk is referred to as the "queen of textiles" for being a fiber with superior strength coupled with excellent drape, luster, and hand. Silk is also one of the costliest textile fibers. Silk is produced from sericulture activities, which are basically agricultural practices associated with silkworm insect rearing. During the process of sericulture, by-products and wastes are produced in addition to the silk fiber. Since silk is associated with a high cost, even the by-products and wastes have been exploited commercially to generate extra income and thus to increase profit.

The supply of waste silk is drawn from the following sources:

- The silkworm, when commencing to spin, emits a dull, lustreless and uneven thread with which it suspends itself from the twigs and leaves of the tree upon which it has been feeding, or the straws provided for it by attendants in the worm-rearing establishments: this first thread is unreliable, and, moreover, is often mixed with straw, leaves and twigs.

- The outside layers of the true cocoon are too coarse and uneven for reeling; and as the worm completes its task of spinning, the thread becomes finer and weaker, so both the extreme outside and inside layers are put aside as waste.

- Pierced cocoons, that is, those from which the moth of the silkworm has emerged-and damaged cocoons.

- During the process of reeling from the cocoon the silk often breaks; and both in finding a true and reliable thread, and in joining the ends, there is unavoidable waste.

- Raw silk skeins are often re-reeled; and in this process part has to be discarded: this being known to the trade as gum-waste. The same term—gum-waste—is applied to "waste" made in the various processes of silk throwing; but manufacturers using threads known technically as organizes and trams call the surplus "manufacturer's waste".

Processing

A silk "throwster" receives the silk in skein form, the thread of which consists of a number of silk fibres wound together to make a certain diameter or size, the separate fibre having actually been spun by the worm. The silk-waste spinner receives the silk in quite a different form: merely the raw material, packed in bales of various sizes and weights, the contents being a much-tangled mass of all lengths of fibre mixed with much foreign matter, such as ends of straws, twigs, leaves, worms and chrysalis. It is the spinner's business to straighten out these fibres, with the aid of machinery, and then to so join them that they become a thread, which is known as spun silk.

All silk produced by the worm is composed of two substances: fibroin, the true thread, and sericin, which is a hard, gummy coating of the fibroin. Before the silk can be manipulated by machinery to any advantage, the gum coating must be removed, really dissolved and washed away. Where the method used in achieving this operation is through fermentation, the product is called schappe. The former, schapping, is the French, Italian and Swiss method, from which the silk when finished is neither so bright nor so good in colour as the discharged silk; but it is very clean and level, and for some purposes essential, as, for instance, in velvet manufacture.

Textile Recycling

Textile recycling is the process by which old clothing and other textiles are recovered for reuse or material recovery. It is the basis for the textile recycling industry. The necessary steps in the textile recycling process involve the donation, collection, sorting and processing of textiles, and then subsequent transportation to end users of used garments, rags or other recovered materials.

The basis for the growing textile recycling industry is, of course, the textile industry itself. The textile industry has evolved into a $1 trillion industry globally, comprising clothing, as well as furniture and mattress material, linens, draperies, cleaning materials, leisure equipment and many other items.

The Urgency to Recycle Textiles

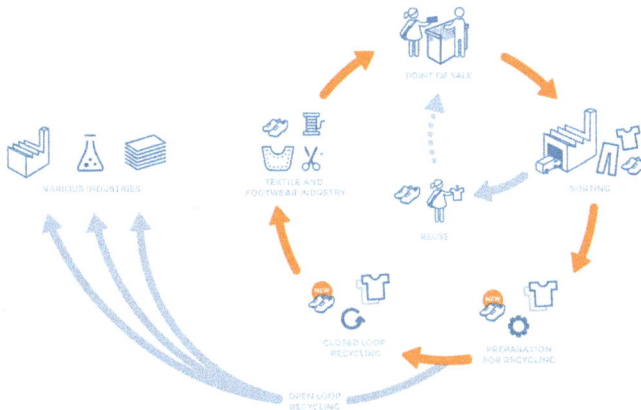

The importance of recycling textiles is increasingly being recognized. Over 80 billion garments are produced annually, worldwide. In 2010, about 5% of the U.S. municipal waste stream was textile scrap, totalling 13.1 million tons. The recovery rate for textiles is still only 15%. As such, textile recycling is a significant challenge to be addressed as we strive to move closer to a zero landfill society.

Once in landfills, natural fibers can take hundreds of years to decompose. They may release methane and CO_2 gas into the atmosphere. Additionally, synthetic textiles are designed not to decompose. In the landfill, they may release toxic substances into groundwater and surrounding soil.

Textile recycling offers the following environmental benefits:

- Decreases landfill space requirements, bearing in mind that synthetic fiber products; do not decompose and that natural fibers may release greenhouse gasses;

- Avoided use of virgin fibers;

- Reduced consumption of energy and water;

- Pollution avoidance;

- Lessened demand for dyes.

For consumers the most common way of recycling textiles is reuse through reselling or donating to charity (Goodwill Industries, Salvation Army, etc.). However, certain communities in the United States have been accepting textiles in curb side pickup since 1990. The textiles must be clean and dry in order to be accepted for recycling. For instance, in 2018 the city of Somerville, MA began a curb side textile recycling program with the city's contractor, Simple Recycling. Each resident receives two pink recycling bags for free for their clothes and other household textiles. The bags are picked up by the contractor and replaced with the same number of bags. Officials in Niles, IL claimed residents diverted over two tons of waste from the landfill with the voluntary textile recycling program in August 2018. The say that the program will help reduce landfill costs as well as add additional revenue from the textiles that will "trickle down to the taxpayer". Additionally, Simple Recycling also operates the program in Niles, IL. They pick up the bags, weigh the donations, and pay Niles $20 per ton. The donations are then sorted and sent to thrift stores, overseas markets, or shredded for insulation and other industrial purposes.

Some companies, such as Patagonia, an outdoor clothing and gear company, accept their product back for recycling. Other companies, like US Again, are for-profit textile recycling companies using collection bins at a variety of sites. Textile recycling equipment plays an important part in the textile recycling industry - Standard and high-efficiency textile recycling equipment is quite important for supporting the textile industry. So far, the most popular and widely accepted clothing recycling bin uses a high safety chutes that are easily opened and closed.

Some textiles can be remade into other pieces of clothing, while damaged textiles are sorted out to make industrial wiping cloths and other items.

Resale

After collection of the textiles, workers sort and separate collected textiles by color, size and quality, it is then packed, baled and sold as good reusable clothing. Shoes are reused by being resold as well. This process not only creates local jobs, it helps stimulate local economy.

Obstacles

If textile re-processors receive wet or soiled clothes, however, these may still end up being disposed of in landfill, as washing and drying facilities are not present at sorting units. This then affects the environment.

Process

Clothing fabric generally consists of composites of cotton (biodegradable material) and synthetic plastics. The textile's composition will affect its durability and method of re-cycling.

Fiber reclamation mills grade incoming material into type and color. The color sorting means no re-dying has to take place, saving energy and pollutants. The textiles are shred-ded into "shoddy" fibers and blended with other selected fibers, depending on the intend-ed end use of the recycled yarn. The blended mixture is carded to clean and mix the fibers and spun ready for weaving or knitting. The fibers can also be compressed for mattress production. Textiles sent to the flocking industry are shredded to make filling material for car insulation, roofing felts, loudspeaker cones, panel linings and furniture padding.

For specialized polyester-based materials the recycling process is significantly differ-ent. The first step is to remove the buttons and zippers; then the garments are cut into small pieces. The shredded fabric is then granulated and formed into small pellets. The pellets are broken down polymerized and turned into polyester chips. The chips are melted and spun into new filament fiber used to make new polyester fabrics.

Some companies are creating new pieces of clothing from scraps of old clothes. By com-bining and making new additions, the eclectic garments are marketed as a type of style.

Sources of Textiles for Recycling

Textiles for recycling are generated from two primary sources. These sources include:

1. Post: consumer, including garments, vehicle upholstery, household items and others.

2. Pre: consumer, including scrap created as a by-product from yarn and fabric manufacture, as well as the post-industrial scrap textiles from other industries.

The donation of old garments is supported by non-profit as well as many corporate programs, including those of Nike and Patagonia.

Wearable and Reused Textiles

According to a U.K. industry source, about 50% of collected textiles are reused, and about 50% are recycled. About 61 percent of recovered wearable clothes are exported to other countries. In some African countries, as many as 80% of people wear used clothing. The issue of sending used clothing to Africa has generated some degree of controversy as to the benefits of such initiatives, where it can have an adverse impact on local textile industries, native dress, and local waste generation.

In Canada, an estimated 10% of charitable contributions are sold by thrift stores, with another 90% of donated fabrics going to textile recyclers. Approximately 35% of donated clothing is made into industrial rags.

Textile Wastewater Treatment

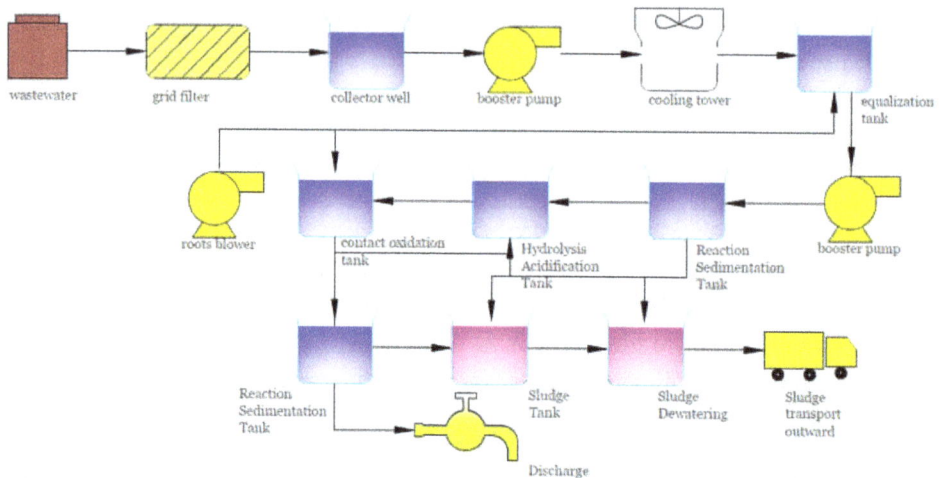

Traditional Method of Treatment

- A treatability study of textile wastewater by traditional method using coagulation by adding polyelectrolytes ((1-2 ppm) at pH (6.7-7.5) and primary sedimentation followed by aeration and final settling gave a good results.

- COD decreased from 1835 to 120 ppm, SS decreased from 960 to 120 ppm and sulphate from 1350 to 125ppm.

- In the full-scale treatment plant filtration is used to improved results by decreasing COD, from 263 to 55 and SS from 295 to 10 and Sulphate from 158 to 100 ppm respectively.

Coagulation and Aeration

The primary treated wastewater from the industrial plant is generally discharged into the public sewer where it is mixed with domestic wastewater.

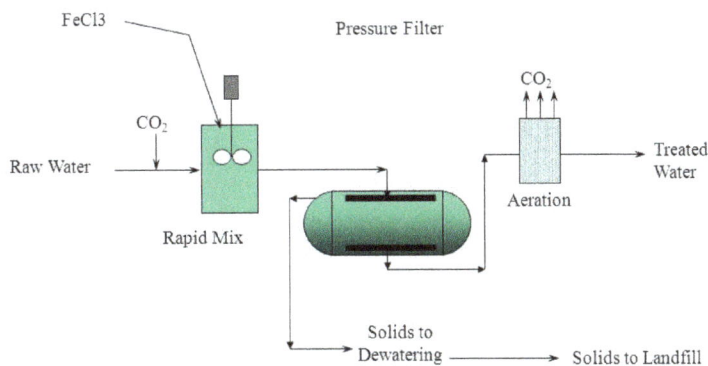

Using coagulation and aeration with sedimentation as secondary or biological treatment improved the effluent quality.

Textile wastewaters have impurities, dissolved colloidal and suspended form, at first are coagulated and precipitated to produce micro-flocks by simply adjusting pH (6.7-7.5) or by adding inorganic or organic coagulating chemical.

Low molecular weight non-ionic polyelectrolytes (1-2 ppm) were used to reduce sludge production when the impurities are in the form of micro-flocks and other suspended solids. The bench scale reactor tanks was operated at HRT of 6 hours without filtration.

Usual Process Outcome

Textile wastewater treatment study has been successfully operated at many textile mills. The simple aerobic treatment in combination with the use of filtration resulted in a significant Suspended solids, COD and color removal. Research was conducted on the development of applicable method of textile wastewater treatment to satisfy the values of SS, BOD, COD, SO 4 and PO 4 removal efficiency showing 76%, 84%, 86%, 92% and 100% respectively without filtration. Removal efficiency improved with filtration for SS, BOD, COD and SO 4 to be 98.3%, 91.2%, 95% and 92% respectively. Using filtration is essential to get treated textile wastewater satisfies permissible limits to be reused.

Technologies for Colour Removal

There are more than 100,000 commercially available dye exist and more than 700000 tonnes per year are produced annually (Pearce et al., 2003, McMullan et al., 2001). Wastewater containing dyes is very difficult to treat, since the dyes are recalcitrant organic molecules, resistant to aerobic digestion, and are stable to light. A synthetic dye in wastewater cannot be efficiently decolorized by traditional methods. This is because of the high cost and disposal problems for treating dye wastewater at large scale in the textile and paper industries. The technologies for colour removal can be divided into three categories: biological, chemical and physical (Robinson et al., 2001). All of them have advantages and drawbacks.

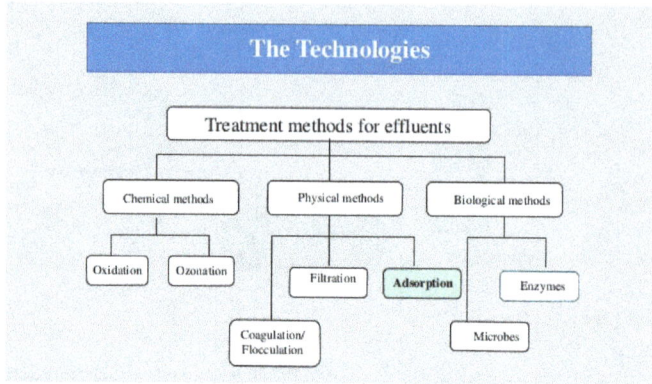

Biological Methods

- Biological treatment is the often the most economical alternatives when compared with other physical and chemical processes. Biodegradation methods such as fungal decolourization, microbial degradation, adsorption by (living or dead) microbial biomass and bioremediation systems are commonly applied to the treatment of industrial effluents because many microorganisms such as bacteria, yeasts, alges and fungi are able to accumulate and degrade different pollutants.

- However, their application is often restricted because of technical constraint.

- Biological treatment requires a large land area and is constrained by sensitivity toward diurnal variation as well as toxicity of some chemicals, and less flexibility in design and operation.

- Further, biological treatment is incapable of obtaining satisfactory colour elimination with current conventional biodegradation processes.

- Moreover, although many organic molecules are degraded, many others are recalcitrant due to their complex chemical structure and synthetic organic origin. In particular, due to their xenobiotic nature, azo dyes are not totally degraded.

Chemical Methods

- Chemical methods include coagulation or flocculation combined with flotation and filtration, precipitation-flocculation with Fe(II)/Ca(OH)2, electroflotation, electrokinetic coagulation, conventional oxidation methods by oxidizing agents (ozone), irradiation or electrochemical processes.

- These chemical techniques are often expensive, and although the dyes are removed, accumulation of concentrated sludge creates a disposal problem.

- There is also the possibility that a secondary pollution problem will arise because of excessive chemical use.

- Recently, other emerging techniques, known as advanced oxidation processes, which are based on the generation of very powerful oxidizing agents such as hydroxyl radicals, have been applied with success for the pollutant degradation.

- Although these methods are efficient for the treatment of waters contaminated with pollutants, they are very costly and commercially unattractive. The high electrical energy demand and the consumption of chemical reagents are common problems.

Physical Methods

- Different physical methods are also widely used, such as membrane – filtration processes (nanofiltration, reverse osmosis, electrodialysis) and adsorption techniques. The major disadvantages of the membrane processes is that they a limited lifetime before membrane fouling occurs and the cost of periodic replacement must thus be included in any analysis of their economic viability. In accordance with the very abundant literature data, liquid-phase adsorption is one of the most popular methods for the removal of pollutants from wastewater since proper design of the adsorption process will produce a high-quality treated effluent.

- This process provides an attractive alternative for the treatment of contaminated waters, especially if the sorbent is inexpensive and does not require an additional pre-treatment step before its application.

- Adsorption is a well-known equilibrium separation process and an effective method for water decontamination applications.

- Adsorption has been found to be superior to other techniques for water re-use in terms of initial cost, flexibility and simplicity of design, ease of operation and insensitivity to toxic pollutants. Decolourisation is a result of two mechanisms: adsorption and ion exchange, and is influenced by many physio-chemical factors, such as, dye/sorbent interaction, sorbent surface area, particle size, temperature, pH, and contact time. Adsorption also does not result in the formation of harmful substance.

References

- Bookwalter, Genevieve. "Clothing, textile recycling program in Niles bringing cash to village as items bypass landfill". chicagotribune.com. Retrieved 2018-08-14.

- "Somerville launches curbside textile recycling program - The Boston Globe". BostonGlobe.com. Retrieved 2018-08-14.

- The-basics-of-recycling-clothing-and-other-textiles-2877780: thebalancesmb.com, Retrieved 17 July 2018

- Mitchell, Kathy. "People, planet, profit focus of Stone Mountain textile recyclers". The Champion Newspaper. Retrieved 4 September 2013.

- Councils "need to understand" importance of textile quality, www.letsrecycle.com, Retrieved 24.11.06

Chapter 6

Other Industrial Wastes and their Treatment

Industries produce a lot of waste such as chemical solvents, sludge, pigments, ash, metal, radioactive wastes etc. These require specialized treatment systems for safe disposal, which are different depending on whether the waste is solid, liquid or radioactive. The aim of this chapter is explore the common industrial waste treatment methods, for photographic waste treatment, pesticide waste treatment, rubber industry wastes treatment, pulp and paper mill wastes treatment, etc.

Photographic Waste Treatment

Major Photochemicals

Photography is very dependent on chemicals. Wastewater from the photographic process contains contaminants such as: hydroquinone, sodium sulfite, silver, mercuric chloride, cadmium, ferrocyanide, acids, and formaldehyde. The types of wastes include: process bath wastes, color developer wastes, bleach, fixer and fixer wastes. All are toxic and highly alkaline. Laws affecting photo processing include the Safe Water Drinking Act and the Toxic Substances Control Act.

Silver Exposure In Biosphere

Silver is a rare but naturally occurring metal deposited as mineral ore, extracted from the main ore, Argentite. It is recovered as a by-product from smelting nickel or copper ores and from platinum & gold deposits. Sources of silver include emissions from smelting operations, manufacture & disposal of photographic /electrical supplies, coal combustion & cloud seeding. Sorption is the main process that controls silver's entry into water & its movement in soils.

Waste Disposaloptions

About 100 million gallons of silver-bearing wastewaters are produced annually. ¬ Materials that leach more than 5 mg/L of silver are classified hazardous waste. Most POTW's can tolerate and treat photo wastewaters - if the volumes and concentrations of contaminants are not too high. ocal sewer authorities regulate concentrations & volumes per day of chemicals released into sewer systems.

Silver Bearing Waste

Silver in photo processing waste is less toxic than "free" silver which can kill aquatic organisms. Silver thiosulfate is the dominant silver compound in photo processing effluents. High concentrations of organic & sulfur-based materials in municipal wastewater treatment systems ensure that any active silver materials are combined into sludge.

Silver-bearing Waste

After silver is discharged into natural waters, it is found tightly-bound in sediments, where it remains and not released into the water column. Silver nitrate can cause skin, eye, and respiratory irritation. Argyria, a condition characterized by bluish-gray pigmentation of the skin, mucous membranes and eyes, is a major effect resulting from long-term chronic exposure to silver.

Onsite Sewage Disposal Septic System

Subsurface disposal systems are designed to manage domestic wastes. Septic systems do not have the ability to properly treat photographic effluents since septic systems operate with anaerobic biological treatment. MDE does not recommend discharging photo processing chemicals into septic systems because these chemicals upset the functioning of the systems and adversely affect nearby underground drinking water sources.

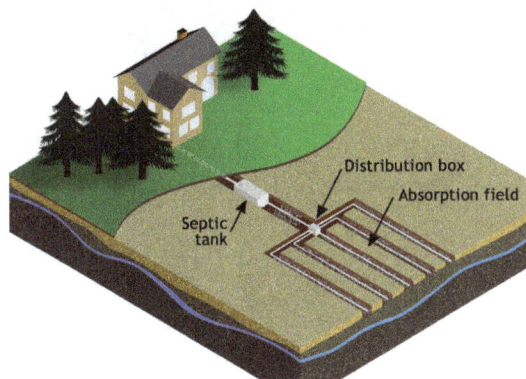

Code of Management Practice

The Code of Management Practice for Silver Dischargers is a voluntary set of recommendations on technology, equipment, & procedures for controlling silver discharges and preventing pollution. The purpose is to develop a consensus among localities and photo-processing silver dischargers to reduce regulatory burdens and costs for municipalities and small businesses. The Code involves the use of recovery and equipment options best suited to a facility, to enhance silver recovery & track the monitoring processes for effectiveness. Silver has a secondary drinking water standard of 0.1 parts per million (ppm).

Silver Recovery

Silver Recovery harvests silver from photo processing solutions. There are several ways to recover silver, depending on the operation size, the concentration of silver in the effluent, and the silver discharge limits for the local POTW. Concentrations of silver in used fixer usually exceed allowable limits for discharge to municipal water systems. The silver recovery process controls laboratory costs and maintains the lab's regulatory compliance.

Types of Silver Recovery

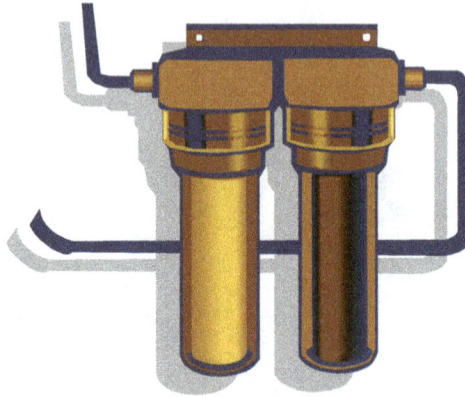

Metallic Replacement involves an active solid metal, such as iron, contacting a solution containing dissolved ions of a less active metal. Electrolytic Recovery applies a direct current across 2 electrodes in a silver-bearing solution. Metallic silver deposits on the cathode. Chemical Precipitation mixes a precipitation agent with silver-bearing waste-water in a batch reaction tank with pH control. Solid particles are formed, settle before filtering, and are sent to a silver refiner.

Silver Recovery from Rinse Water

Ion Exchange is the reversible exchange of ions between a solid resin and a liquid. It can recover up to 98% of silver. Reverse Osmosis has the wastewater flow under pressure over the surface of a selectively permeable membrane. Water molecules pass through the membrane, and other constituents are left behind, recovering 90 % of the silver thiosulfate.

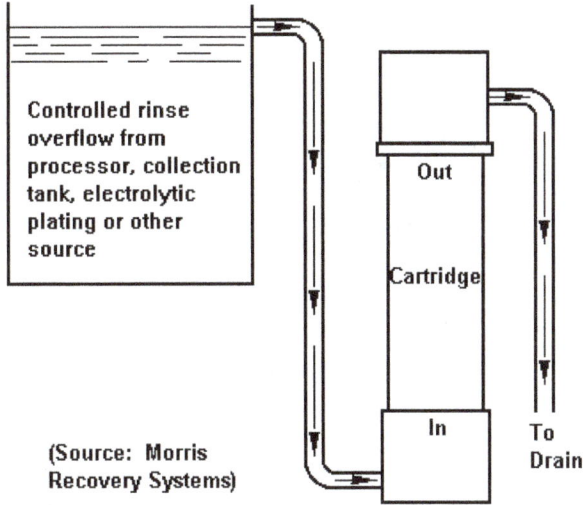

Controlled rinse
overflow from
processor, collection
tank, electrolytic
plating or other
source

Out

Cartridge

In

To
Drain

(Source: Morris
Recovery Systems)

Silver Recovery Options

Developer/fixer disposal can be handled through an off-site silver reclamation facility, licensed to accept hazmats. Operate your own silver recovery unit. This unit must be operated under certain regulations. Ensure that the silver concentrations in your system are acceptable to the local sewer system.

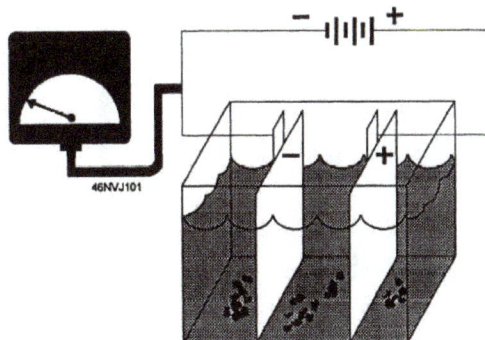

Best Management Techniques

Seal all floor drains connected to the sewer or storm drains by production area. Any solutions touched by developer must be put in hazmat. receptacle for pickup by licensed hauler. Use squeegees to wipe excess from films and papers. This saves $$ & chemical quantities. Install secondary containment around all machines. Keep waste fluid segregated for reuse, recycling or trt.

Best Management Techniques

Floating lids on replenisher tanks reduce oxidation, evaporation & contamination from dirt. Replace highly toxic developers, such as catechol, chlorquinol, and pyrogallo with less toxic ones, such as phenidone. Donate un emulsified inks to school or public agencies.

Best Management Techniques

Clean up spills at once. Use absorbent materials to confine any fluids. Replace solvent-based plate-making systems with water-bases ones. Reduce amount of waste rinse water by using counter current rinse tanks. Install automatic ink levelers to keep ink fountains at optimal level for good print quality in large web presses.

Photochemical Recycling

Choose inks & cleaning solutions that are non-toxic. Avoid halogenated compounds, petroleum-based, or phenol cleaners. Recycle spent fixer, solvents, waste ink. Strip "goldenrod" from negatives and used metal plates & accumulate for pickup by a licensed hauler. Accumulate "chromoliths" for recycling.

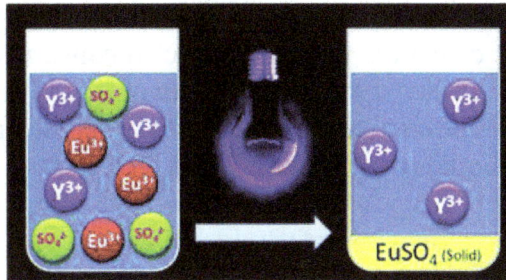

Safe Substitutes

Use soybean, walnut, or vegetable oil-based inks for lithography printing. Water-based inks can be used for screen printing. Reduce concentration of Isopropyl Alcohol (IPA) with a fountain solution with low IPA or switch to low-VOC substitutes. Use soap solutions when possible. Solvents should be used only for cleaning inks & oils. Some specially made blanket washes & acetic acid-based solvents, with less hazmats, are now available. Some small solvent recovery systems are on market, able to accommodate many medium-large printers.

Photorecycling News

Wash less minilabs use a stabilizer instead of wash water, recovering silver effluent and discharging it only to a municipal secondary treatment system. Digital cameras now rival conventional photography. They recycle by reducing size & run on fewer batteries. Some companies remove lead from lenses, cadmium from sensors, & mercury from displays. The disposable camera is the most recycled consumer product. Polystyrene covers & viewfinders are grinded down into new camera components and lens acrylic is made into toothbrushes. One company extracts 99% of the silver and other toxic heavy metals from used photo liquids to produce a liquid fertilizer.

Pesticide Waste Treatment

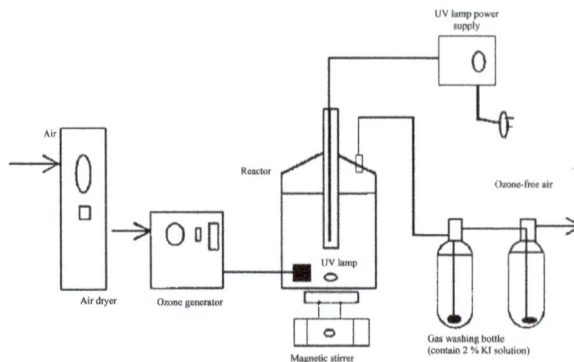

Pesticides played a vital role in the economic production of wide ranges of vegetable, fruit, cereal, forage, fibre and oil crops which now constitute a large part of successful agricultural industry in many countries. They lower crop losses, increase revenue to farmers from the additional marketable yield obtained with their use and thus lower the cost of production per unit output. Other benefits include: 1) reduced uncertainty of crop loss from pests, 2) increased profit to farm input suppliers (machinery, fertilizer, chemicals and seed companies) from increased sale, 3) benefit to consumers through decreased price of raw foods or improved quality of food products and 4) benefit to society as whole (farmers, consumers, farm suppliers, food processors) from increased employment opportunities and expanded export.

Pesticide	Pest to be controlled
Insecticide	Insects
Herbicide	Undesirable plants
Rodenticide	Rats, mice and other rodents
Nematicide	Nematodes
Fungicide	Fungal diseases
Acasicide	Mites and spiders
Bactericide	Bacteria

Table: Common types of pesticides.

Pesticide	Consumption (10^6 kg)	Expenditures (10^6 \$)
Herbicides	955	15,512
Insecticide	405	11,158
Fungicide	235	9216
Other*	775	3557
Total	2370	39,443

*Other includes mematicide, fumigants and other miscellaneous.

Table: 2007 worldwide consumption and expenditures of pesticide active ingredients.

Country	Application Rate (kg/ha)
Costa Rica	51.2
Colombia	16.2
Japan	12.0
Netherlands	9.4
Korea	6.6
Ecuador	6.0
Portugal	5.3
France	4.6
Greece	2.8
Uruguay	2.7

Table: Top 10 countries applying pesticide at higher rates in 2000.

Company	Country	Sales (10^6 $)	Market Share (%)
Bayer	Germany	7458	19.0
Syngenta	Switzerland	7285	18.5
BASF	Germany	4297	10.9
Dow Agro Science	USA	3779	9.6
Monsanto	USA	3599	9.1
Du Pont	USA	2369	6.0
Makhteshim Agan	Israel	1895	4.8
Nufarm	Australia	1470	3.7
Sumitomo Chemical	Japan	1209	3.0
Aystra Life Science	Japan	1035	2.6
Total	N/A	34,396	87.2

Sales are in millions of dollars; Total Worldwide Sales = 39443.

Table: Top 10 pesticide companies in 2007

of food products. The benefit/cost ratio vary from 4 to 33 (for every dollar spent on pesticide farmers receive an additional $4 - 33 in revenue) depending upon crop rotation and year.

After pesticides are applied to the target areas, pesticide residues remain in containers and application equipment. These residues are removed from applicators by rinsing with water which results in the formation of a toxic wastewater that can adversely affect people, pets, livestock and wildlife. The resulting ecological impact of unsafe disposal of pesticides can be severe depending on the type of pesticide and the amount contained in the wastewater. The phenomenon of bio magnification of some pesticides has resulted in reproductive failure of some fish species and egg shell thinning of birds such as peregrine falcons, sparrow hawk and eagle owls. Pesticide toxicity to humans includes skin and eye irritation and skin cancer. Therefore, care must be exercised in the application, disposal and treatment of pesticides. Currently, disposal of pesticide wastewater is carried out by several methods including: 1) land cultivation, 2) dumping in soil pits, in ditches, in lagoons, on land, and in extreme cases in sewers and streams near the rinsing operation, 3) use of evaporation pond and 4) land filling. These methods of disposal are totally unsafe. The surface run off will reach streams, rivers and lakes and the infiltration of the wastewater into the local soil will eventually end up in the ground water. The treatment methods currently used for pesticide containing wastewater include: 1) incineration, 2) chemical treatment, 3) physical treatment and 4) biological treatment. These treatment methods either require land or are expensive and suffer from variability of effectiveness. Thus, the development and selection of safe, on farm disposal/treatment technique for agricultural pesticides is paramount.

Pesticide Disposal Methods

The methods for the disposal of low level pesticides include: land cultivation, disposal pits, evaporation ponds and landfills. There are three types of disposal pits: soil pit, plastic pit and concrete pit.

Land Cultivation

In this method, excavated contaminated soil is spread out in a thin layer on uncontaminated soil in order to allow for natural chemical and biological processes to transform and degrade the contaminants. Soil contains microbes (fungi, algae and bacteria) capable of metabolizing pesticides.

Method	Description	Advantages	Disadvantages
Land Cultivation	Place liquid wastes in plow zone of soil for subsequent weathering	On-site use Simple technology	Land requirements Possible runoff and leaching Slow and variable decomposition Restricted vegetation
Disposal Pits	Place liquid wastes in pits containing soil and open to air for subsequent weathering	On-site use Simple technology Secure containment	Slow decomposition Limited lifetime of pit Effectiveness varies with climate
Evaporation Ponds	Place liquid wastes in lined ponds open to air for subsequent weathering	On-site use Simple technology Secure containment	Slow decomposition Limited lifetime of pond effectiveness Varies with climate
Landfills	Burial of wastes in soil	Generally available Complete removal	Land requirements High transportation costs Possible runoff and leaching

Table: Disposal methods of pesticide containing wastewater.

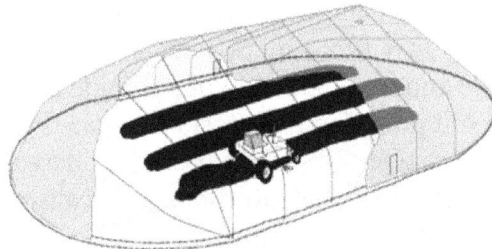

Figure: Land cultivation of contaminated soil.

The ability of bacteria to metabolize pesticides has been well documented by several researchers. Bhadhade reported that soil bacteria was capable of degrading 83% - 93% of the organo-phosphorouspesticide monocrotophos. Ohshiro reported a 96% reduction in isoxathion from the organo-phosphouruspestiside by bacteria isolated from turf green soil. Kearney reported that soil microbes were capable of degrading 90% of the alachlor pesticide within 30 - 40 days. Tang and You reported that the triazophos bacteria was capable of degrading 33.1% - 95.8% of pesticides in soil.

Racke and Coats reported that after soil has been treated with a pesticide a few times its microorganisms build up a need for that pesticide which results in fairly rapid degradation of any additional applications. Schoen and Winterlin stated that natural soil degradation is effective when low concentrations of the pesticide are present, but with high concentrations of pesticide it becomes much more difficult to degrade. Felsot stated that land cultivation is only effective for compounds that can be biotransofrmed or biominerlized by soil microbes. Somasundaram reported that the ability of soil microbes to degrade certain pesticides is affected by pesticide toxicity to soil microbes that are responsible for the degradation. Felsot stated that land cultivation is effective if the pesticide is degraded at the same or faster rate than it is applied to the field. Felsot noted that land cultivation can be enhanced by the addition of organic amendments such as sewage sludge.

Soil Pit

A primary method for disposing of liquid pesticide waste is by dumping it in an unlined soil evaporation pit, usually $15 \times 15 \times 1$ m. Schoen and Winterlin reported that factors such as chemical structure and concentration of pesticide play a major role in the degradation of pesticides in soil pits. Gan and Koskinen stated that the dissipation of the pesticide decreases as the concentrations of pesticide increases. Dzantor and Felsot and Gan noted that high pesticide concentrations may cause microbial toxicity which would inhibit the degradation of the pesticide.

Several researchers noted that the prolonged dissipation of pesticides opens a window for runoff and leaching, especially at higher pesticide concentrations. Gan reported that 50% dissipation of the alachlor pesticide in soil, at concentrations of 4 and 4 300 mg/kg took approximately 2 and 52 weeks, respectively. Gan noted that atrazine pesticide took approximately 4 and 24 weeks to be dissipated to half the concentration of 7 and 6400 mg/kg in soil, respectively. Schoen and Winterlin noted that captan, trifluralin and diazinon at concentrations of 100 mg/kg took 1 - 2, 116 - 189 and 77 - 160 weeks to dissipate to half the concentration while captan, trifluralin and diazinon at concentrations of 1000 mg/kg took 30 - 48, 168 - 544 and 77 - 160 weeks to reach 50% disappearance in soil, respectively.

Plastic Lined Pit

This method for disposal of pesticide waste requires proper selection of the site to avoid leaching and runoff.

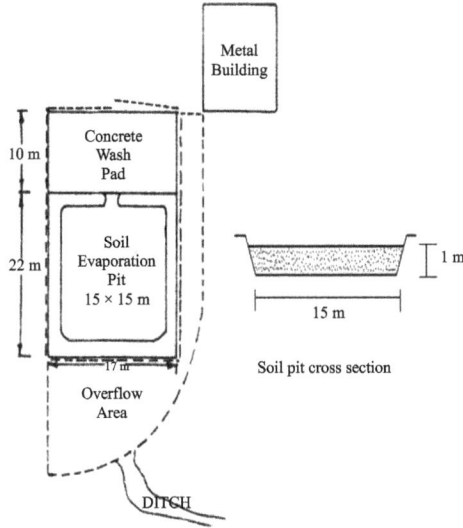

Figure: A soil pit for disposal of pesticide water

The site should be in an area where there is no danger of contaminating dwellings groundwater sources and surface water used for crop and livestock production. The pit should be on a levelled ground with a depth of 0.5 - 1 m covered with a plastic liner and a layer of soil is laid on top of the liner. The pit should be open to the atmosphere in order to allow for water evaporation into the atmosphere. A roof cover will prevent the water level from raising due to rain or snow. The wastewater is pumped into the pit for pesticide biodegradation by soil microbes. Hall reported that the presence of microbes in the soil water mixture in plastic lined pits was responsible for the degradation of pesticide and no accumulation of pesticide was noted in the pits. Junk and Richard evaluated the effectiveness of 90,000 L polyethylene lined disposal pit with over 150 kg of 25 different types of pesticides for over 2 years and concluded that this method was in fact effective for disposal of pesticide waste with insignificant release to air and water surroundings.

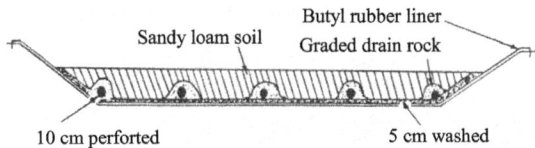

Figure: A plastic lined evaporation pit for disposal of pesticide wastewater

Figure illustrates a cross section view of a simple small scale plastic pit used to dispose of pesticide waste. It consists of a plastic drum with a length and width of 75 × 55 cm, respectively. Inside the drum is a mixture of 15 kg of soil and 60 L of water that was used to treat pesticide waste which was introduced into the system through the inlet. Junk eused 56 plastic containers filled with 15 kg of soil and 60 L of water to test the degradation of alachlor, atrazine, triflualin, 2,4-D ester, carbaryl and parathion and found this system not suitable for atrazine but was effective and very rapid for 2,4-D and carbayl.

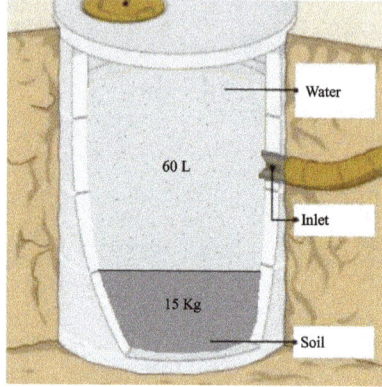

Figure: A cross section of a plastic pit for disposal of pesticide wastewater

They concluded that: 1) the plastic container provided satisfactory containment for most common pesticides, 2) soil was a satisfactory source for microorganisms, 3) aeration and buffers had questionable value, 4) half-life concept for degradation was not applicable and 5) sampling from small disposal sites was a problem.

Concrete Pit

Similar to the plastic lined pit, the concrete pit should be on levelled ground with a depth of 0.5 to 1 m, a length of 8 - 10 m and a width of 3.5 m and reinforced with 0.20 m thick concrete walls. The pit consists of a top and bottom layer of gravel that is 4 cm in diameter with the middle layer consisting of topsoil. The pit should also have a cover to prevent rise in water level from rain or snow but remain open to the atmosphere in order to allow for water evaporation.

Johnson and Hartman tested the microbiological activity in a concrete pit and concluded that the degradation process in the pit was effective and no long-term accumulation of pesticide was present. Junk and Richard tested the effectiveness of 30,000 L concrete disposal pit with over 50 kg of 40 different types of pesticides for 8 years and concluded that this method was in fact effective for disposal of pesticide waste with insignificant release to air and water surroundings. Hall tested the effectiveness of an open concrete disposal pit for the degradation of 45 pesticides over five months and concluded that the biodegradation of the pesticides was successfully accomplished and the pit did not leak or pollute the air, but the system was too large and complicated for most farms.

Evaporation Beds

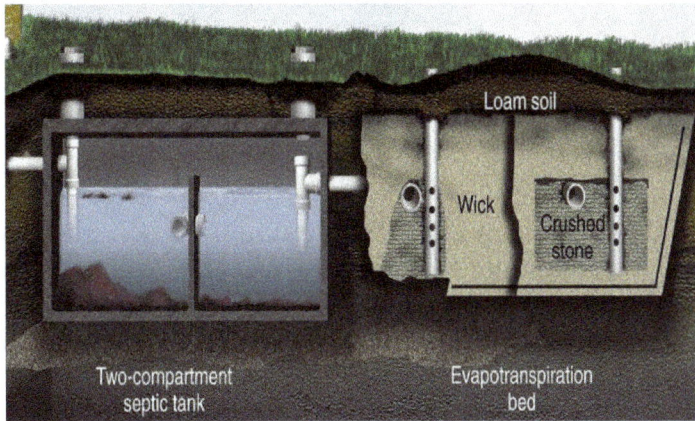

Lined evaporation beds are used for the disposal of pesticide wastewater. Leach lines underneath the soil surface supply the beds with the pesticide residues from washing equipment. The pesticides rise to the beds surface where they are degraded through photochemical, chemical and biological actions and are distributed via air vapour. Some of the beds have hydrated lime incorporated into the soil in order to aid in the degradation of certain pesticides. A medium scale disposal system of this type costs up to $50,000 to construct.

Hodapp and Winterlin reported a reduction in the diazinon pesticide of 62.54% using lined evaporation bed without lime and a degradation of 77.75% with lime. They also reported an ethyl parathion reduction of 69.83% using lime treatment in the beds and a reduction of 45.45% without the use of lime. Winterlin tested ten (6 × 12 × 1 m) lined (with a butyl rubber membrane liner and 36 cm of sandy loam soil) evaporation beds to determine their pesticide decay effectiveness. Pesticide rainsate was introduced through subsurface tiles in limited amounts and the effects of geography, climate and lime application were examined. The method appeared to be beneficial for disposal of some pesticides but not all. Over 100 pesticides were tested, but only 46 were actually detected.

This method for the disposal of pesticide containing wastewaters is advantageous because the beds are economical, little maintenance is required, do not build up high levels of pesticides and are effective in degrading as well as containing the pesticides without excessive exposure through air vapour. It is considered an economical, on-site method of disposal which requires only annual monitoring. The disadvantages appear to be the development of a high concentration of residue in the top layer of the soil and the difficulty in acquiring a representative sample.

Land Filling

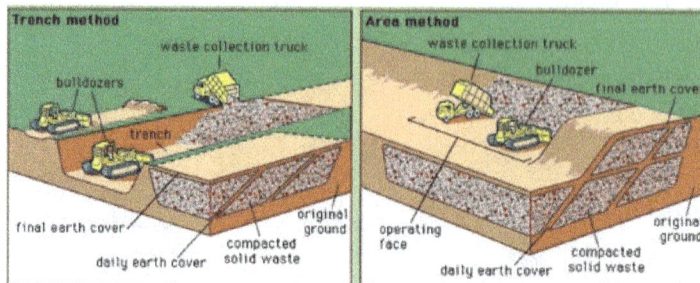

Landfills are sites that dispose of waste by burial into the soil where microorganisms are used to change the composition of the toxic elements. Landfills for pesticides are equipped with drying pits containing soil to provide the microbes needed to break down the pesticide components into non harmful elements. A nearby sump for the propose of draining and rinsing the containers that have not been fully emptied or rinsed.

Munnecke reported that soil bacteria were capable of hydrolyzing ethyl parathion found in pesticide container residues within 16 h. Johnson and Lavy reported that carbofuran, thiobencarb and triclopyr buried in degrading containers dissipated to 50% of the initial concentration with the first 94 days or less, while benomyl took 179 - 1020 d before 50% dissipation. They also noted that the rates of dissipation decreased with an increase in soil depth.

Yasuhara detected 190 compounds in landfill leachates in Japan. Williams reported that the pesticide mecoprop is found in landfill leachate because it is resistant to anaerobic degradation. Christensen noted the presence of the pesticide bentazon, N,N-Diethyltoluamide and mecoprop in landfill leachate because of their persistence to anaerobic landfill conditions. Alloway and Ayres noted the presence of the pesticides atrazine and simazine in landfill leachate.

Pesticide Treatment Methods

The pesticides treatment methods include: 1) thermal treatment, 2) chemical treatments, 3) physical

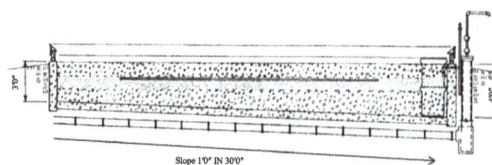

Figure: A cross section of concrete pit for disposal of pesticide wastewater

Figure: Evaporation beds for disposal of pesticide wastewater

Figure: A landfill for disposal of pesticide wastes

Figure: The landfill leachate collection pit

treatments and 4) biological treatments. Thermal treatments include incineration and open burning. Chemical treatments include ozonation/UV radiation, Fenton oxidation, hydrolysis and KPEG. Physical treatments are based on absorption using activated carbon, inorganic and organic materials. Biological treatments include composting, phytoremediation and bio augmentation.

Incineration

Pesticide Incineration is a high temperature oxidation process where the pesticide is converted into inorganic gases (water vapour, CO_2, volatile acids, particles and metal oxides) and ash. Incineration of pesticide should be operated at temperatures higher than 1000°C so that the pesticide can be treated within the first 2 seconds. At such temperatures, smoke production is nil and the generated combustion gases are similar to those generated by wood burning. Temperatures lower than 1000°C can also be used as long as the incineration time of the pesticide does not exceed 2 seconds. However, lower temperatures tend to produce toxic intermediate products.

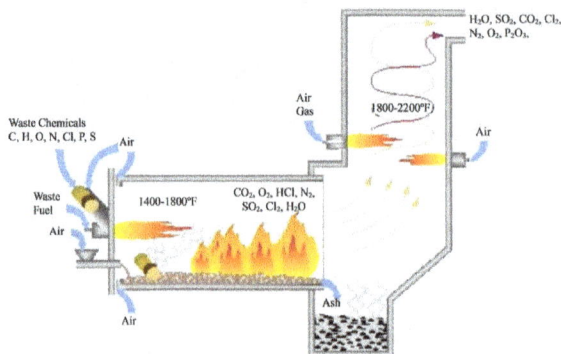

Kennedy noted change in the combustion efficiency over the temperature range of 600°C - 1000°C. Ferguson and Wilkinson reported that incineration has 99.99% destruction efficiency at temperatures of 1000°C and a retention time of 2 s in the combustion zone. Steverson reported a 99% destruction efficiency for 16 currently used insecticides and herbicides at temperatures ranging from 200°C to 700°C. Linak reported an incineration efficiency of greater than 99.99% for dinoseb. Ahling and Wiberger noted that the incineration of fenitrothion and malathion at temperatures lower than 600°C gave emissions of 1% - 2% of the pesticide amount added and temperatures above 700°C would be required to achieve safe destruction and emission levels.

Method	Description	Advantages	Disadvantages
Thermal	Controlled combustion of either liquid waste or concentrated residue	Destructive Rapid No by-products	High costs Complex Not useful for some chemical
Chemical	Chemical destruction through use of oxidative, reductive, hydrolytic or catalytic reagents	Destructive Rapid	High costs Complex Variable effectiveness
Physical	Removal of chemicals from wastewater by adsorption and/settling	Rapid Possible on-site use	No destruction involved By-products for disposal
Biological	Use of micro-organisms to destroy chemicals	Destructive	High costs Susceptible to shock Relatively slow Variable effectiveness

Table: Current treatment methods of pesticide containing wastewater.

Incinerators capable of achieving high levels of destruction are equipped with a combustion chamber, an afterburner, scrubbers and electrostatic filters. Ferguson and Wilkinson reported the following performance standards for incinerating hazardous wastes: 1) the incinerator must achieve a destruction and removal efficiency greater than 99.99% for each of the chemicals present in the waste feed, 2) HCl emissions must not exceed 1.8 kg/h or 1% of the HCl in the stack gas prior to entering any pollution control equipment and 3) the particulate matter emitted must not exceed 180 mg/DSCM when corrected to 7.0% O_2. The advantages of incineration include: 1) effectiveness in degrading chlorinated organics, 2) destruction efficiency of 99.99% and 3) setup at locations next to plants generating the waste. The disadvantages of incineration technology include: 1) need for sophisticated equipment 2) production of cyanide in the off gas during the incineration of organonitrogen pesticides, 3) too costly and complex, 4) it is intended for centralized large scale disposal and 5) not recommended for inorganic pesticides.

Open Burning

This method combusts pesticides and pesticide waste containers by piling up empty paper and plastic containers and setting them on fire. Although this method is inexpensive and convenient, it is hazardous to workers, plants and animals. It is prohibited in some cases by the Regional Air Quality regulations in the US. It emits gases, smoke and fumes into the atmosphere as well as toxic residues that are left in the containers.

Adebona noted several products of incomplete combustion, polyaromatic hydrocarbons and low levels of dioxins in open burning tests on 22.7 kg insecticide bags. Oberacker noted that after burning bags containing phorate, 2% of the phorate was released into the air and 0.5% remained in the solid residues. Felsot reported that bags containing atrazine released 13% of the remaining product into the air while 25% remained as residue. Such results indicate that the temperatures for complete combustion were not reached or were not maintained long enough in order to obtain destruction efficiencies of 99.99% or greater.

Ozonation/UV Radiation

The use of ozone and UV radiation to enhance the oxidation of aromatic compounds was investigated by several researchers. Ozonation is more effective treatment method in the presence of UV light because it conform hydrogen radicals which are very effective oxidizing agents. The benefits of this process are its mobility, ease of operation and rapid effects. The disadvantages are its high energy consumption and initial equipment cost.

Kuo used a UV/ozonation system consisting of a medium-pressure mercury vapor lamp with a water cooling jacket and an ozone generator. The lamp power consumption was 150 W and was capable of 14.3 W output (at 3.0 mW/cm² at a distance of 9 cm). The O_3 was pumped at a rate of 400 mg O_3/hr/L solution. A solution of 2% KI was used for absorbing the residual ozone from the reactor.

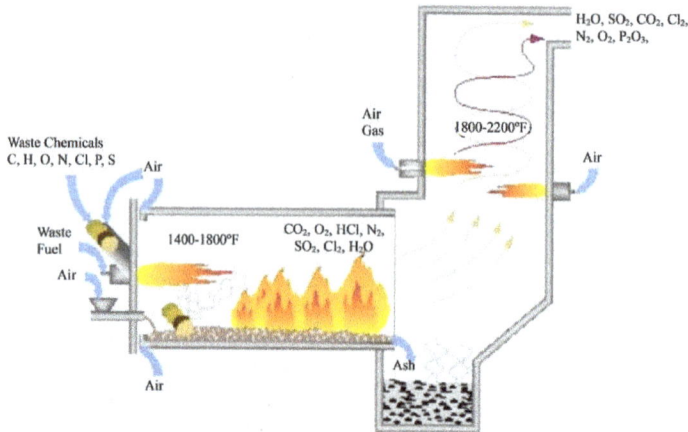

Figure: An incinerator for pesticide wastes.

Figure: Open burning of pesticide wastes.

Somlich noted that irradiation of the alachlor pesticide achieved de-chlorination of the compound, while ozonation works to oxidize the compound into several intermediate products. Under Ultraviolet irradiation, the photon absorption by the carbonyl present in the compound is then followed by the loss of the chlorine. The pesticide degradation reactions that take place under UV/ozonation are as follow:

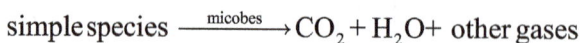

$$pesticide + O_3 \xrightarrow[H_2O]{UV\ light} CO_2 + H_2O + simple\ species\ gases$$

$$simple\ species \xrightarrow{micobes} CO_2 + H_2O + other\ gases$$

Kearney monitored the degradation of alachlor using a UV/O_3 system by measuring the concentration of the 14 CO_2 released. Under UV radiation, the alachlor pesticide was completely depleted from the water with the presence of oxygen within 25 minutes while it and took 50 minutes before it was fully depleted with ozone alone.

Fenton Oxidation

The Fenton process can be used as part of an oxidative system to treat and degrade pesticides. It consists of hydrogen peroxide (H_2O_2) and iron salts at low pHs. The iron salts act as a catalyst, increasing the effectiveness of the H_2O_2 by forming highly reductive hydroxyl radicals. The radicals are capable of oxidizing other species that are present in the solution as follows:

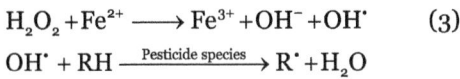

$$H_2O_2 + Fe^{2+} \longrightarrow Fe^{3+} + OH^- + OH^{\cdot} \qquad (3)$$
$$OH^{\cdot} + RH \xrightarrow{\text{Pesticide species}} R^{\cdot} + H_2O$$

Hydroxyl radicals are very powerful oxidizing agents with a 2.33 V oxidative potential. The rate of degradation of organic pollutants is strongly accelerated by UV irradiation. The photolysis of the Fe^{3+} complexes allows the regeneration of Fe^{2+} thus allowing the reaction to proceed much quicker in the presence of H_2O_2. The advantages of this method for pesticide treatment are: low cost, ease of operation, simplicity and the wide range of temperature that can be used.

Fallmann noted a 72% reduction in 100 ppm total organic carbon solution using 23 mL of hydrogen peroxide and a reaction time of 124 minutes in a photo assisted Fenton process. Larson reported that in the presence of ferric perchlorate and a mercury lamp, the atrazine pesticide had a half-life of less than 2 minutes compared to 1500 minutes when iron salt was not present.

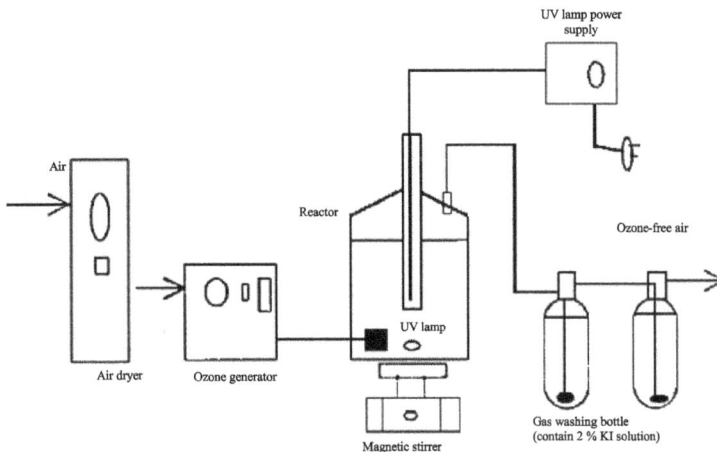

Figure: A UV/O3 system for the treatment of pesticide wastewater.

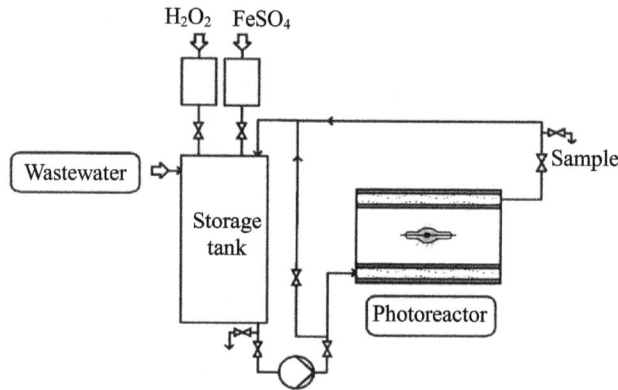

Figure: A photo assisted Fenton system for treatment of pesticide wastewater.

Huston and Pignatello noted a half-life of less than 10 minutes for the captanpesticide using UV assisted Fenton reagent at a pH of 2.8. Pignatello and Sun reported a half-life of 2 minutes for methyl parathion using UV assisted Fenton reagent. Doong and Chang reported a half-life of less than 10 minutes for alachlor pesticide under photo assisted Fenton reagent at a pH of 2.8.

Hydrolysis

This method for pesticide treatment works by hydrolyzing the ester linkages found in pesticide compounds, including pyrethroids, carbamates, organophosphates and acetaniledes. These compounds can be hydrolyzed in solutions with high pH levels. Desmarchelier used calcium hydroxide for ester hydrolysis and found it to be a safer alternative to sodium and potassium hyoxides for the hydrolysis of fenitrothion pesticide. Lee noted that under basic conditions, sodium perborate was more effective in the hydrolysis of organophosorus than sodium hydroxide, because the peroxide anion released from sodium perborate is much more reactive to organ phosphorus than the hydroxyl ion. Qian noted an enhancement in the hydrolysis process of mevinphos, diazinon, methyl parathion, malathion and parathion in lake water (10 mg/L) using sodium perborate at pH of 9.88. However, with the presence of soil, the reaction was noted to be significantly slower and the concentration of perborate had to be increased by four folds.

Metal oxide and divalent metal ions have been noted for their ability to catalyze the hydrolysis of organphosorus insecticides. Smolen and Stone reported that the phophorothionate insecticides (chlorpyrifos-methyl, zinophos, diazinon, parathion-methyl and runnel) and phosphorooxonates (chlorpyrifos-methyl oxon and paraoxon) were most effectively catalyzed by Copper (II). The downside of catalysis using metal oxides is the formation of products with significant toxicity. Badawi and Ahmed noted that the hydrolysis of the pesticides diazinon, cypermethrin and carbaryl was effective and accelerated by the addition of a copper (II) ion complex.

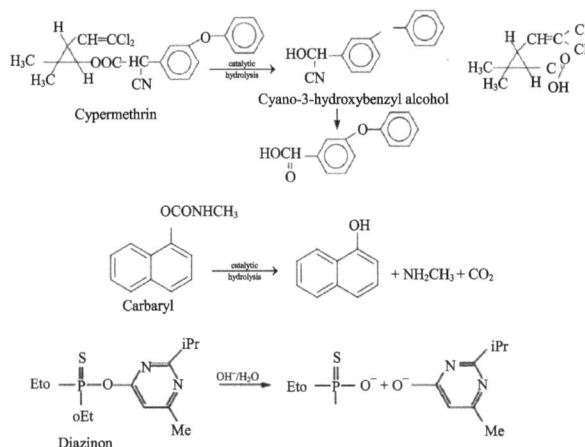

Figure: Hydrolysis of cypermethrin, carbaryl and diazinon pesticides.

KPEG

Potassium polyethylene glycol ether (KPEG) is capable of destroying chlorinated pesticides. Chlorinated hydrocarbons and cyclodienes are resistant to degradation by hydrolysis. Dechlorinating these pesticides with KPEG would then enable their biodegradation through land treatment processes. KPEG was found to be capable in dechlorinating polychlorinated biphenyls (PCBs) in soil and solvents. In older formulations of phenoxy herbicide, KPEG was found to be capable of degrading dioxins and dibenzofurans . The reaction that takes place consists of a nucleophilic substitution and a phase transfer at the carbon-halogen bond as illustrated by the following equations.

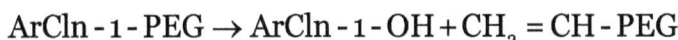

$$PEG + KOH \rightarrow KPEG + H_2O$$

$$KPEG + ArCln \rightarrow ArCln\text{-}1\text{-}PEG + KCI$$

$$ArCln\text{-}1\text{-}PEG \rightarrow ArCln\text{-}1\text{-}OH + CH_2 = CH\text{-}PEG$$

where:

PEG = polyethylene glycol monomethyl ether Ar = aromatic nucleus Taylor reported that the vessel for the KPEG reaction consists of a 55-gal drum (surrounded with heat tape capable of maintaining the temperature at $70°C - 85°C$) and an electric motor with a mixer. With the reagents KOH and PEG, this vessel was capable of degrading 98% of phenoxy herbicide waste. The generated waste contained in the drums can remain there, eliminating the need for transfer into another container. Vapor emitted from the reaction drums are condensed in a water drum, the remaining condensables are traped in the scrubber containing sodium hypochlorite solution. Vapors are then passed through an activated carbon absorbent and as well as a molecular sieve.

The materials and chemicals needed for the KPEG process are easy to find. The disadvantage of the KPEG process are 1) high clay content, acidity and high natural organic matter interferes with KPEG reaction and 2) its not recommended for large waste volumes with concentrations above 5% for chlorinated contaminants. If necessary, emissions can be controlled by construction of a vent system with scrubber and absorbent.

Inorganic Absorbents

Pesticide adsorption can be performed using anionic clays (layered double hydroxides) which are simple to prepare, hydrotalcite, which occurs in nature may also be used as a layered double hydroxide (LDH). A variety of compounds can be formed by changing the cation metal. In order for a material to be considered as a good adsorbent it must possess the following properties: 1) a granular structure, 2) insoluble in water, 3) chemical stability and 4) have a high mechanical strength. Figure is multi-functional gravity filter which can be used for various water treatment methods by employing various adsorbent media.

Niwas reported that styrene supported zirconium (IV) tungstoophosphates was successful in adsorbing the pesticide phosphamidan. Inacio noted that the inorganic adsorbent Mg3AlCl was capable of adsorbing the MCPA herbicide within 30 - 45 minutes at room temperature. Boussahel noted a removal efficiency in ayanaz in and atrazine of 85% - 90% using $CaCl_2$ or $CaSO_4$. Bojemueller reported that the pesticide metolachlor can be adsorbed by bentonites and the adsorption efficiency can be doubled by increasing the temperature. Li noted that the pesticide glyphosate was adsorbed on the external surface of MgAl-LDH at low concentrations, while at high glyphosate concentrations an inner layer ion exchange occurred.

Organic Absorbents

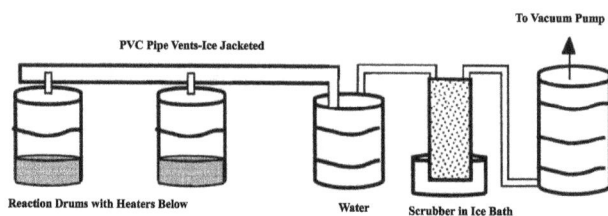

Figure: A KPEG process with vent system.

Various organic materials can be used as good adsorbents for pesticide removal. Ahmaruzzaman and Gupta reported that rice husk is insoluble in water, possesses an irregular granular structure and has a high mechanical strength and chemical stability that make it a good adsorbent. Chowdhury noted that treated rice husk was capable of removing 89% - 97% of malachite green pesticide. Akhtar investigated the adsorption potential of selected agricultural waste materials (rice, barn, bagasse fly ash from sugarcane and rice husk) for the pesticide removal of methyl parathion from wastewater and reported pesticide removal efficiencies in the range of 70% - 90% within 90 minutes. Memon et al. reported that thermally treated watermelon peels were capable of removing 99% of the methyl parathion pesticide. Al hattab and Ghaly reported a captan removal efficiency of 99.2% and 98.5% using hay and soybean plant residues, respectively.

Figure: Filters used to treat pesticide wastewater through adsorption.

Activated Carbon

Carbon adsorption treatment method for pesticide containing wastewater is used in the pesticide manufacturing industry as well as in pesticide cleanup. The activated carbon system consists of a prefilter made up of sand or an alum flocculation chamber with a carbon filter. Dennis and Kobylinski reported on a Carbolator system which uses a suspended bed of carbon packed in bags of floating porous polyethylene in order to avoid clogging. The water was continuously recirculated through the carbon filters by directing it back into the waste holding tank.

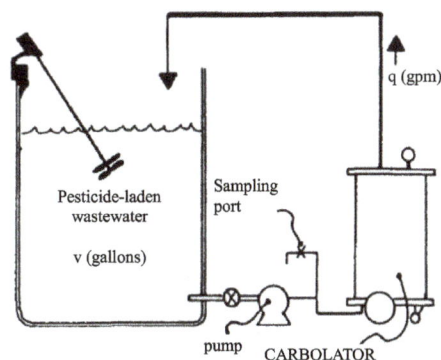

Figure: Recirculation through activated carbon.

Felsot reported that rinsewater containing malathion, propoxur, chlorpyifos, diaxinon and dimethoate were all removed to nondetectable levels using the Carbolator. This process reduced the amount of waste generated by several magnitudes through efficiently absorbing pesticides form the water. Kobylinski used a Carbulator 35B to remove baygon, dimethoate, diazinon, runnel, malathion, dursban and 2,4-D and found that the higher the molecular weight of the compound the more favourable the effect of adsorption by activated carbon. Similar findings were also reported by other researchers.

Honeycutt reported that a waste stream containing 100 ppm chlorophenols was reduced to 1 ppm using activated carbon. Giusti reported a carbon activated adsorption of 3.6% and 98.5% for methanol (molecular weight of 32 g/mol) and 2-ethyl hexanol (molecular weight of 130.2 g/mol), respectively. Sarkar reported an adsorbent efficiency of 98% - 99% for the removal of the isoproturon pesticide using powdered activated charcoal. Gupta reported an adsorption efficiency of 70% - 80% using activated charcoal for removing pesticides. Word and Getzen reported that a decrease in pH increased the adsorption of aromatic acid compounds due to enhancement of carbon surface properties.

The activated carbon is very effective in removing pesticides and it does not require extensive monitoring. The disadvantages include: 1) the need for a skilled chemist for field testing, 2) the high cost and 3) this process is only capable of adsorbing solutions with concentrations of less than 1000 ppm.

Composting

This treatment method relies primarily on microbial activity and aeration efficiency. Microorganisms that are naturally occurring in the materials increase significantly in numbers and begin to decompose biodegradable compounds which results in the release of carbon dioxide as well as the production of metabolic heat, causing the temperature of the compost to rise to 60°C - 70°C. As the compost temperature increases, three successions of microbes occur: psycrophilis, mesophilis and thermophilis.

Racke and Frink reported a complete degradation of carbaryl during the composting of sewage sludge. Petruska achieved a complete degradation of diazinon pesticide using dairy manure compost. Rose and Mercer reported a 100% degradation of parathion insecticide in cannery wastes. Singh reported a degradation efficiency of 96.03% for the endoslufin pesticide after 4 weeks, using composted soil with a moisture level of 38%. Al hattab and Ghaly achieved a captan removal efficiency of 92.4% in the first four days using hay compost.

Several researchers stated that polyhalogenated hydrocarbons, used in pesticides, can be metabolized under anaerobic conditions. However, other researchers noted that pesticides may largely persist unchanged during the composting process. Muller and Korte noted little to no degradation of the aldrin, dieldrin and monolinuron during the composting of sewage waste sludge. Strom noted the presence of chlordanein finished compost from various US municipalities.

Phytoremediation

In this method plants are used to contain and remove harmfull environmental contaminants as shown in figure below. Kruger reported a degradation efficiency in atrazine of 65% after 9 weeks in soil where Kochi sp. was planted. Coats and Anderson reported that degradation of atrazine, metrolachlor and triflualin was enhanced in soils where the Kochi sp. plant grows. Olette reported that the aquatic plants L. minor, C. aquatic and E. Canadensis were capable of removing 2.5% - 50% of dimethomorph and flazasulfuron present in the water. Buyanovsky noted that the fungi rhizosphere was capable of degrading carbofuran by using it as its carbon source. Gordon noted that 95% of trichloroethylene was removed from wastewater by hybrid polar trees during growing season. Stearman noted that in constructed wetlands, cells with plants were capable of removing 77.1% and 82.4% of simazine and metolachlor, respectively, while cells without plants were only capable of removing 64.3% and 63.2%, respectively. Wang reported that in the first 20 days of plant growth, oilseed rape seedlings were capable of removing 20% of chlorpyrifos pesticide.

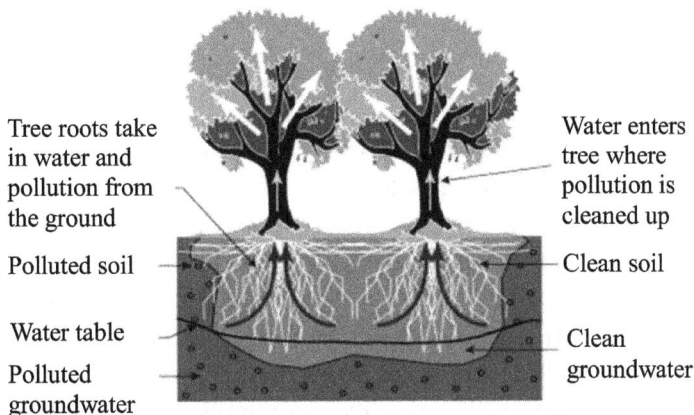

Figure: Phytoremediation to clean up pesticide contaminated grounds.

Bioaugmentation

This method uses isolated microbes for the degradation of pesticides. The pesticides are quickly metabolized and converted to products with a lower toxicity under aerobic and anaerobic conditions. Some pesticides are capable of being degraded by certain bacterial strains as their sole nutrient or carbon source. Such microbial reactions are known as mineralizing because of the large amount of carbon dioxide released during metabolism.

Dichlorodiphenyltrichloroethane, normally resistant to biodegradation, has been reported to have been metaboilized in an anaerobic microbial culture, followed by an aerobic one. Bhadhade reported that bacteria isolated form soil was capable of degrading 83% - 93% of the organophosphorous pesticide, monocrotophos. Ohshiro reported a 96% reduction in isoxathion from the organophosphourus pestiside by bacteria isolated from turf green soil. Tang and You, reported that the triazophos bacteria is capable of degrading 33.1% to 95.8% of pesticides.

The down side of such a method is the establishment of the microbes in the presence of other microbial populations present in the contaminated soil. Acea noted that the population of the introduced bacteria may be reduced due to susceptibility to predation or starvation.

Pulp and Paper Mill Wastes Treatment

Activated Sludge Process

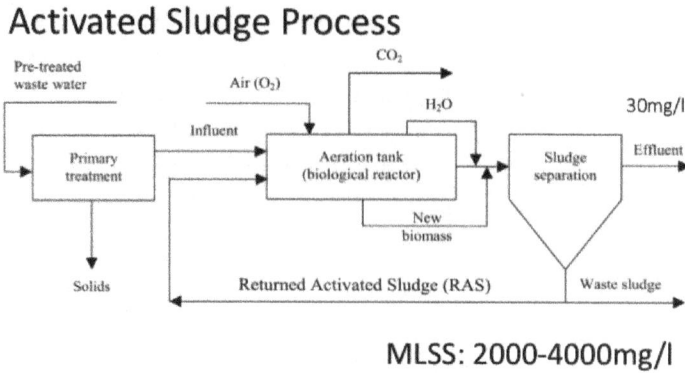

MLSS: 2000-4000mg/l

The pulp and paper mill industry is an intensive consumer of water and natural resources (wood), discharging a variety of liquid, gaseous and solid wastes to the environment. Since the 1970s, a growing awareness of the effects of pulp and paper wastes in the ambience had prompted water and energy consumption levels and the loads of toxic compounds discharge to reduce. One of the most important implemented changes in this regard was made within the mill, wherein chlorine was completely substituted by, that is, chlorine dioxide as the bleaching chemical agent. Another major issue was the implementation of secondary biological treatments. The wastewater composition and hence the effluent treatment efficiencies and characteristics of the discharges are strongly dependent on the technology applied and the raw materials. In the last 25 years, however, the global distribution of pulp producers has significantly changed and so have the species of wood used. Eucalyptus pulp production, for example, is becoming a leader in the hardwood pulp market; Brazil went from being a pulp consumer to a world leader in hardwood pulp production, and since 2008, it has been the fourth largest pulp producer in the world.

Wood Pulp Market

Cellulose pulp is the main raw material in the production of different types of paper and paperboard. It is also used as the absorbent material in diapers and other sanitary products.

The global pulp market has changed intensely in recent years. A few decades ago, this industry was characterized as national character as a supply industry inputs for domestic production of paper and paperboard. Globalization has led to increased competitiveness in the international market, as new players have emerged both at the level of producers and consumers. Within the latter, the appearance of China and India have strongly modified cellulose demand worldwide.

Figure graphically shows the evolution of world's production of wood pulp between 1979 and 2013 according to the data published by FAO.

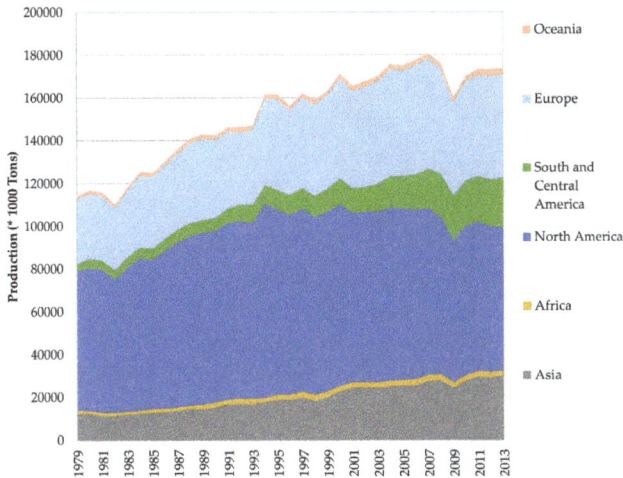

Figure: World pulp production 1979–2013.

It is clearly shown that the world's wood pulp production increased to about 50% in this period, from 120 million tons in 1979 to nearly 180 million tons in 2013. Part of this growth can be explained by the explosive increase in production in non-traditional wood pulp producing regions such as Asia and South America. The main producing regions are still North America with 38% and Europe with 28%, even though in 2013, Asia produced 17% of the wood pulp and South America about 13%.

Wood pulp grades are categorized according to the pulping process, which can be classified as mechanical, semi-chemical and chemical pulps. In a mechanical process, logs or wood chips are mechanically grinded by abrasive action. In a chemical cooking process, a significant part of the wood components (mainly lignin) is chemically dissolved to obtain a solid compound with high cellulose fibre content. There are two main methods of chemical pulping: (1) sulphite pulping and (2) sulphate (kraft) pulping. The first process—sulphite cooking process—uses aqueous sulphur dioxide (SO_2) and a base of calcium, sodium, magnesium or ammonium. The kraft process uses a treatment comprising a mixture of sodium hydroxide and sodium sulphide, known as white liquor, at a high pressure and temperature. The semi-chemical pulping process combines chemical and mechanical methods, where wood chips are first softened or partially cooked with chemicals and then mechanically pulped.

Figures illustrate the different kinds of pulp produced in 1979 and 2013.

The rise in wood pulp production is due to an increase in chemical pulp production, as the production of mechanical pulp has declined in the same period. Mechanical pulping has the advantage of converting up to 95% of dry weight wood into pulp, although considerable amounts of energy are required to do so. The pulp obtained produces a highly opaque paper with good printability, but the physical properties are inferior than chemical pulps and yellowing when exposed to light. Moreover, mechanical pulps are mainly produced from softwood.

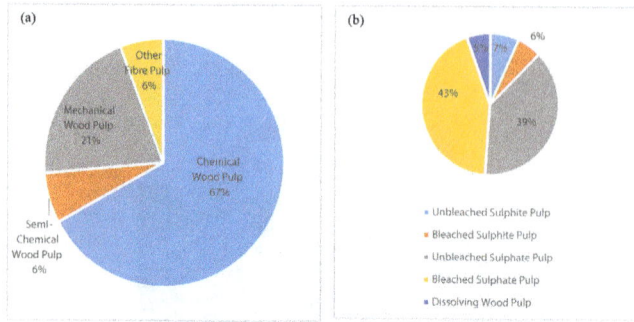

Figure: (a) World pulp production by type of pulp in 1979;
(b) different kinds of chemical pulps produced in 1979.

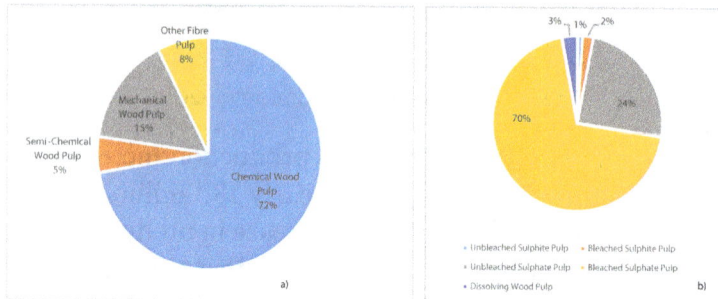

Figure: (a) World pulp production by type of pulp in 2013;
(b) different kind of chemical pulps produced in 2013.

There are significant changes in the production of chemical pulp. The use of sulphite cooking process in pulp production compared to kraft pulping technology decreased steadily, from 60% in 1925 to 20% in 1967 and 9.2% in 1979 to only 2.4% in 2013. The superiority of kraft pulping process is explained by the following facts: (1) all wooden materials including low-quality wood can be used as raw material; (2) superior fibre strength of pulp compared to other chemical pulping methods; (3) more simple chemical and energy recovery process; (4) scale of economy of kraft methods prevents competition and (5) low environmental risks in modern mills.

A second classification considers the type of wood used by distinguishing softwood or long fibre (produced mainly from pine and spruce) from hardwood or short fibre (produced from eucalyptus, birch, poplar, etc.). A gradual move from softwood to hardwood can be observed. In 2013, 56% of bleached kraft pulp was produced with long-fibre wood (softwood), while the remaining 44% was produced with short-fibre wood (hardwood). In 1980, the production capacity of bleached kraft pulp corresponded to 63% of softwood pulp. The entry into the market of non-traditional producing countries such as Brazil, Indonesia, Spain and Portugal, significantly increased the production of hardwood pulp. Eucalyptus bleached pulp production is rapidly increasing (from 8 million tons in 2003 to nearly 15 million in 2015), and eucalyptus wood is thus considered to be the most important raw material of hardwood bleached market pulp in the world.

As kraft pulping is by far the most common process used these days, this chapter will focus in the wastewaters generated in this process.

Main Processes Description

Mechanical Pulping

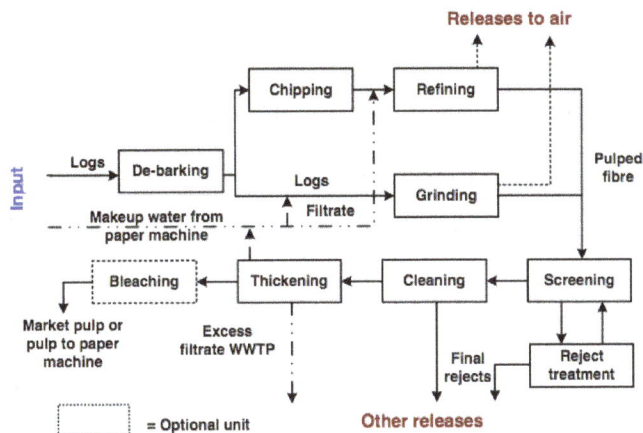

The oldest method of mechanical pulping is the ground wood process. In this process, round logs are forced against a rotating pulp stone (revolving at peripheral speeds of 1000–1200 m/ min), under specified conditions of pressure and temperature. Atmospheric grinding, pressure grinding and thermo-grinding could be done according to the applied temperature and pressure. In all of them, the temperature levels obtained from the heat applied or from rubbing the logs on the stone soften and break down the fibres structure; and cracks the fibres from the wood matrix.

Another common method is the refiner mechanical pulping (RMP). The wood chips are pulled between two rotating disks. Among them, thermo mechanical pulping operates like RMP, but under higher temperature and pressure. The high temperature and pressure levels soften the lignin even more than frictional heat, making fibres separation easier. Thermo mechanical pulp is stronger than refined mechanical pulp, and still retains the high-yield and cost-effectiveness of mechanical pulps.

Chemical Pulping

Sulphite Pulping

Sulphite process is very versatile, and covers the entire pH range, achieving high fibre flexibility in pulp yields and properties. The cooking process involves the use of aqueous sulphur dioxide (SO_2) and a base: calcium, sodium, magnesium or ammonium. Sulphite pulping was developed in the second half of the nineteenth century and for several decades, the calcium acid sulphite process was the most common method.

However, since 1950, the utilization of bases other than calcium has been a major development. The specific base used will determine the process's chemical and energy recovery system and water use. The use of the relatively cheap calcium base has become obsolete because the cooking chemicals cannot be recovered. Magnesium and sodium bases allow chemical recovery, and magnesium bases are currently the dominant choice in sulphite pulping process.

Kraft Pulping

In kraft pulping, white liquor, containing mainly active chemicals—sodium hydroxide and sodium sulphide—is used for cooking the chips at a high temperature (150–170°C) and pressure. Approximately, half of the wood composition degrades and dissolves during cooking. The spent cooking liquor (black liquor) contains reaction products of lignin and hemicelluloses, and is concentrated and burned in a recovery boiler that recovers the cooking chemicals and generates energy. The smelt is dissolved into water to form green liquor (mostly sodium carbonate and sodium sulphide), which then reacts with lime to convert the sodium carbonate into sodium hydroxide regenerating the white liquor. After cooking and washing, a brown pulp (brown stock pulp) is obtained. Printing, writing and tissue papers require the pulp to be bleached which removes the excess lignin and chromophores to produce a "white" pulp.

Background of Pulp Mill Effluents: Environmental Fate and Effects

The pulp and paper industry consumes enormous amounts of water and natural resources and is also one of the largest effluents generators. Before the 1970s, wastewaters from the pulp and paper mills were normally discharged directly to the rivers or lakes, without any treatment or even a rough primary treatment. The high organic loads and solid content in the effluents affected the aquatic ecosystem in several ways such as localized damage to the benthic community, oxygen depletion in large areas and numerous changes in fish reproduction and physiology. In the 1980s, studies in Scandinavia, along the Baltic Coast and the Gulf of Bothnia, showed alterations in fish reproduction and increase of diseases and parasites. Studies conducted in USA and Canada in the beginning of the 1990s, under the Environmental Effects Monitoring (EEM) program, revealed delayed sexual maturity, smaller gonads, changes in fish reproduction and depression in secondary sexual characteristics in species living downstream of pulp and paper mills discharges.

From the end of the 1970s until now, the main concern regarding effluents is the formation of chlorinated compounds in bleaching plants. In 1985, 2,3,7,8-tetrachlorodibenzo-p-dioxin (TCDD) was discovered in the pulp mill effluents, which led to a general concern over the formation of chlorinated organic matter in chlorine bleaching. Consequently, the use of chlorine in the bleach plants gradually decreased until it was completely substituted with chlorine dioxide. In many countries, the environmental control authorities set strict restrictions on the discharges of chlorinated organics, measured as absorbable organic halogen (AOX), into the aquatic environment. In 1992, the Swedish Environmental Protection Agency limited organ chlorines emissions to 1.5 kg AOX/t of pulp and in 1995, Finland's official limit was set at 1.4 kg AOX/t of pulp.

Several authors reported that with the replacement of chlorine with chlorine dioxide, the effluent quality improved in AOX levels and the elimination of detectable amount of dioxins, polychlorinated compounds and chloroform.

The European Integrated Pollution and Prevention Control has created reference documents (BREF) that set the Best Available Techniques (BAT) for several industrial sectors. The pulp and paper industry has a very defined set of operations to be especially applied in the new mills. Similarly, the International Finance Corporation among others has defined directives that could be required to give financial support for the construction of new mills. For kraft pulp, the most important guidelines are listed in Table.

Dry debarking of wood
Extended modified cooking to a low kappa number (batch or continuous)
Systems for collection and recycling of temporary and accidental discharges from process water spills
Closed screening
Efficient washing of the pulp ahead of the bleaching
Oxygen delignification ahead of the bleach plant
Elemental chlorine free (ECF) or total chlorine free (TCF) bleaching
Removal of hexenuronic acids by mild hydrolysis at the beginning of the bleaching process, for hardwood pulp, especially eucalyptus
Partial closure of the bleach plant combined with increased evaporation
Sufficient and balanced volumes of pulp storage, broke storage and white water storage tanks to avoid or reduce process water discharges
Recycling of wastewater, with or without simultaneous recovery of fibres
Separation of contaminated and non-contaminated (clean) wastewaters
Biological secondary wastewater treatment

Table: Main BAT guidelines from IFC and/or IPPC Bureau regarding wastewater load minimization in bleached kraft pulp mills.

Mechanical Pulping: Wastewater Characteristics

Figure: shows a block diagram of the main part of the mechanical pulp production indicating the sources of emissions to the water from a pulp mill.

Table: shows the specific water consumption and loads before wastewater treatment from the mechanical pulping.

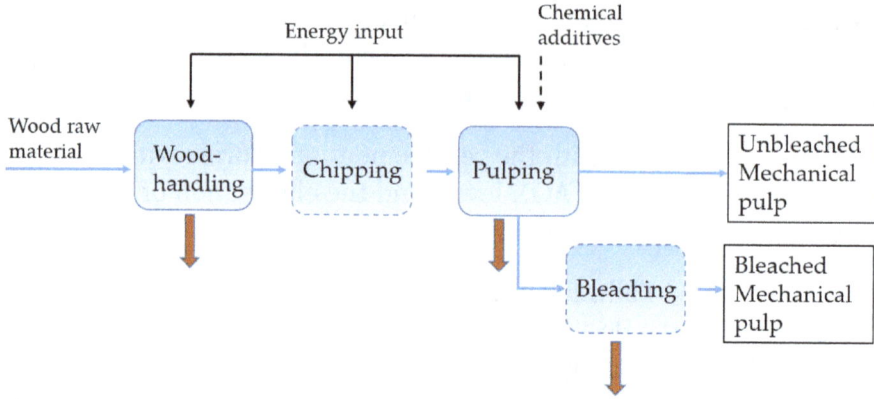

Figure: Main unit operations of the mechanical pulping. Light brown arrows indicate wastewater sources.

Pulping process	BOD$_5$ (kg/ADt)	COD (kg/ADt)	Nitrogen (kg/ADt)	Phosphorous (kg/ADt)
GW	8.5–10	20–30	80–100	20–25
PGW	10–13	30–50	90–110	20–30
RMP	10–15	40–60	90–110	20–30
TMP	13–22	50–80	100–130	30–40

BOD5: Biochemical Oxygen Demand; COD: Chemical Oxygen Demand; GW: ground wood pulping; PGW: pressurized ground wood pulping; RMP: refined mechanical pulping; TMP: thermo mechanical pulping; ADt: air dry tone (10% water and 90% oven-dry pulp).

Table: Specific water consumption, organic and nutrient loads before wastewater treatment from the mechanical pulping.

Kraft Pulping: Wastewater Characteristics

Process Description and Emissions to Water

A kraft pulp mill can be divided into four main parts: (1) raw material handling; (2) pulping line with an almost closed chemical and energy recovery system; (3) bleaching with an open water system and (4) the external wastewater treatment system. Figure shows the emissions sources to water from a kraft pulp mill.

Table: shows the typical figures for the parameters in different sectors of a kraft pulp mill.

Data on current discharges to water (after wastewater treatment) expressed as loads based on available data from kraft pulp mills within the European Union are given in table. Figure presents a comparison of the discharges to water of different existing mills with the performance of the new mills in South America that are processing eucalyptus wood and applying the best available techniques (according to the European IPPC Bureau and the IFC Guidelines.

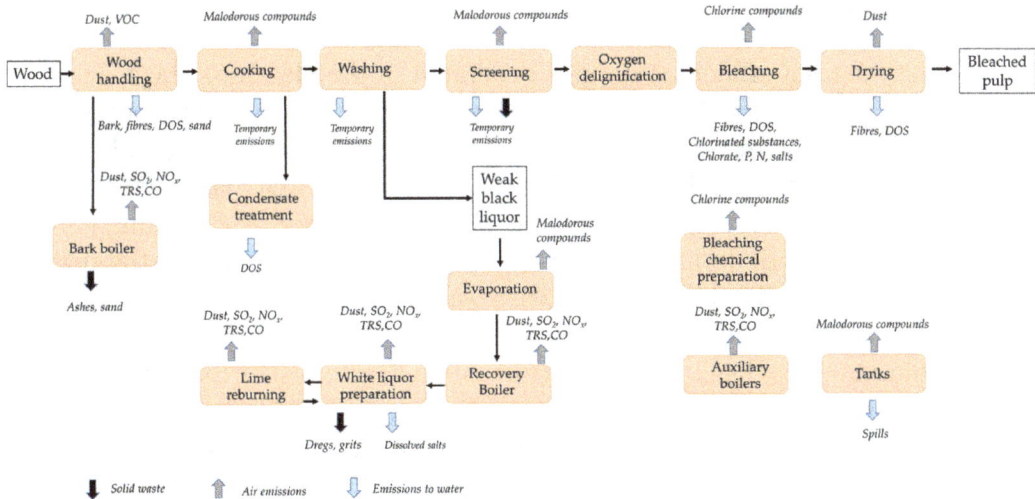

Figure: Main unit operations of kraft pulping. DOS: dissolved organic substances.

Department	Flow	TSS	BOD	AOX	COD	P	N
Debarking	2.5	4	2	0	5	20	0. 2
Washing and screening	0.5	3	1	0	2	1	0.015
Bleaching	31	2	10	1.2	35	47	0.075
Condensates	1	0	1	0	3	0	0
Others	3	4	4	0	10	7	0.002
Total	38	13	18	1.2	55	75	300
Flow in m3 /Adt, TSS, BOD, AOX, COD and Nitrogen in kg/ADt. Phosphorous in g/ADt							

Table: Sources of effluents and effluents loads from kraft pulp mill.

	Flow	COD	AOX	TSS	Total P[1]	Total nitrogen
Unbleached pulp	14–82	1.2–23	–	0.02–3	0–0.05	0.01–1.0
Bleached pulp	20–94	5–202 7.5–423	0–0.3	0.015–7	0.003–0.11	0.01–0.6
Flow in m3 /ADt, COD, BOD5 , AOX, TSS, nitrogen and phosphorous in kg/ADt.						
1 Eucalyptus strands contain higher levels of phosphorus compared to other forest species used for pulp production. The average level discharged with the effluent is up to 0.12 kg total-P/ADt.						
2 Emissions from eucalyptus pulp mills.						
3 Emissions from other hardwood (no eucalyptus) and softwood.						

Table: Reported annual average discharges from kraft pulp mills within the EU

Bleaching Effluent

Up to 85% of the total effluent volume is generated in the bleaching stage. Therefore, this part of the mill is broadly studied in order to minimize the effluent organic loads (especially the organ chlorines loads) without impacting the pulp yield and brightness. Effluent loadings depend on the production process and the raw materials. The degree of delignification of the unbleached pulp, the bleaching process, the washing loss, type of wood, final brightness desired, chemical and water consumption and the degree of plant closure are important indicators of wastewater characteristics. To this end, kappa number is an important mill control parameter. The kappa number quantifies by a redox reaction to the amount of lignin (or the delignification degree) still in the pulp. The higher the kappa number, the higher the lignin content in the pulp. The low lignin amounts to be removed during bleaching, decreases the utilization of bleaching chemicals, which consequently reduces the load to the wastewater treatment. However, if the kappa number were to decrease too much during the cooking then the pulp yield and physical properties will be considerably low. Table provides performance data of the different processes.

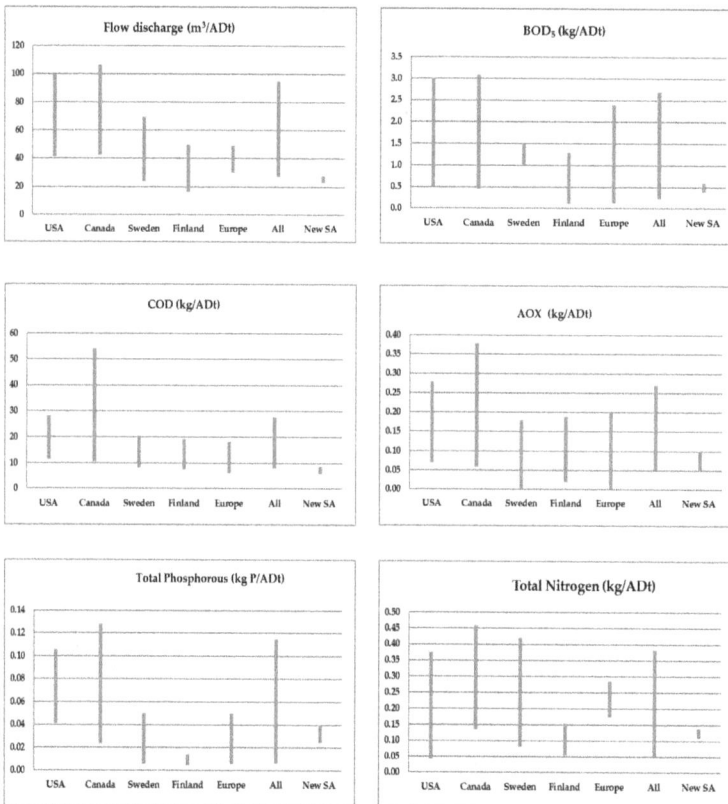

Figure: South America new mills performance compared with mills in North America and Europe. The vertical bars depicted in the graphs correspond to the 10th percentile to 90th percentile range. The column "All" corresponds to the average of the values reported.

Delignification technologies	Kappa for hardwood	Kappa for softwood	Calculated COD load (kg/t) from the bleach plant	
			Hardwood	Softwood
Conventional cooking	14–22	30–35	28–44	60–70
Conventional cooking + oxygen delignification	13–15	18–20	26–30	36–40
Extended/modified cooking	14–16	18–22	28–32	36–44
Extended cooking + oxygen delignification	8–10	8–12	16–20	16–24

Table: Kappa number currently achieved with different delignification technologies and comparison of the calculated effluent COD without considering the washing losses.

The effluents from kraft pulp bleaching constitute varying quantities of organic and inorganic substances. The organic typically represents one-third of the dissolved material while the inorganics comprise two-thirds. The solid matter includes mainly fibres, pieces of fibres and the additives used in bleaching. The dissolved organic matter is composed of various species derived from the raw material and formed in the pulping and bleaching process (residual lignin, hemicelluloses and extractives).

Wood material impact on the values of the effluents parameters can be assessed by comparing the figures for bleaching effluents derived from softwood and hardwood pulp. The former has higher COD and colour content than those of hardwood pulp. The compounds responsible for colour are lignin fragments of high molecular weight (HMW), which represents low biodegradability in the biological treatment. Research has compared effluents from softwood and eucalyptus pulps through AOX, COD, BOD_5 and colour behaviour of the different kinds of pulp production (conventional bleached pulps and oxygen deligni-fied bleached pulps). According to the findings, softwood and eucalyptus effluents have the same trend in AOX levels. For both conventional pulps, the AOX levels were higher than the corresponding oxygen delignified pulps. Furthermore, as it mentioned earlier, the total COD levels are dependent on the initial kappa numbers. The COD compositions of eucalyptus and softwood effluents are significantly different, where the effluents from the eucalyptus pulps are more biodegradable. The compounds forming the kappa number in softwood and hardwood (especially eucalyptus) differ as well: in softwood, the kappa number mainly representative of lignin, whereas in eucalyptus, the hexenuronic acids (HexA) are a large contributor. In this regard, the most common way to remove the hexenuronic acids is in the early bleaching stages through hot acid hydrolysis (A) and hot chlorine dioxide bleaching (D_H) technologies.

Chemical Composition of the Wastewater

The two main types of bleaching methods in use are elemental chlorine free (ECF), when no molecular or gaseous chlorine is dosed in the bleaching, and totally chlorine free (TCF) bleaching. ECF is dominating the bleached chemical pulp market. In 2012,

ECF pulp production reached approximately 93% of bleached kraft pulp's world market share. TCF production has declined a little over the last 10 years.

Owing to the differences between both the bleaching technologies and chemical composition of the bleaching effluents, it is necessary to study in order to predict and understand the environmental impact associated, and consequently to develop the most suitable treatment that decreases effluent loads and toxicity. A significant number of studies pertaining to the chemical composition of bleaching effluents have been published. Several authors have worked in identifying the chemical compounds in filtrates. More than 500 organic compounds have been identified in bleaching effluents so far. Most compounds identified in bleaching effluents are derived from lignin or other wood components, such as extractives or carbohydrates.

The most important difference, when comparing softwood effluents with the eucalyptus effluents, is the higher lignin content in the former and the hexenuronic acid content in the latter.

Lignin degradation products were commonly considered as the major precursors of chlorinated compounds. However, the presence of monochlorinated compounds derived from glucuronxylans were identified to be the major components of chlorine dioxide bleaching filtrates of eucalyptus kraft pulps.

Other important compounds found in the effluents are wood-derived components: resin acids, fatty acids, phytosterols and retene. Lipophilic hardwood extractives consist of a complex mixture of compounds such as sterols, long chain aliphatic acids and alcohols, waxes, glycerides and sterol esters. If high amounts of these compounds are found in kraft mill effluent, their origin is frequently the spills of black liquor and soap or black liquor transported with the pulp.

Molecular Weight Distributions

Several authors have worked in determining the molecular weight distribution of the components in the effluents. The importance of determining the molecular weight distribution comes from the fact that significant removal in the biological treatment system is achieved from the low molecular weight (LMW) material. Evidence of this is the increment in the proportion of organic compounds with high molecular weight after biological treatment. Improvements in the removal of high molecular weight material would lead to greater efficiency and improve the effluent quality. Traditionally, the separation between low molecular weight (LMW) and high molecular weight (HMW) is done at 1000 Da. Bleach kraft mill effluents have an extended molecular weight distribution; from diverse kinds of monomeric compounds to large and complex molecules with molecular weights between 10,000 and 30,000 g/mol. The molecular weight distribution depends on the raw material and the bleaching process used. For example, the average molecular weight of organic matter in hardwood kraft pulp effluents is lower than the corresponding softwood effluents.

The molecular weight fractions in the bleaching filtrates of oxygen delignified eucalyptus pulps were studied. The HMW fraction contributed to approximately 40% of the total effluent load of COD both in softwood and hardwood ECF bleached pulps production, and about 30–40% to TCF bleached pulps effluents. Additionally, the most remarkable differences between softwood- and hardwood-derived effluents are in the aromatic region. The aromatic lignin-derived structures such as syringyl and guaiacyl units are not important structural elements in HMW effluent materials from ECF bleaching of oxygen delignified hardwood kraft pulps, but are important in softwood HMW effluents. Similarly, the results show that all HMW effluents contained carbohydrates. The carbohydrates found in the examined HMW could have had oligosaccharides, polysaccharides or both present in the effluent, either in dissolved or colloidal form. As can be expected, the HMW hardwood kraft pulps fraction contained more carbohydrates (mainly xylan) than the corresponding samples from softwood kraft pulps. Concerning the presence of carboxylic acids, the HMW samples showed high levels of these groups. They were formed due to the oxidation of lignin structures in the bleaching process.

Regarding the low molecular weight (LMW) compounds, it can be broadly classified into three main classes: acids, phenolic compounds and neutral compounds. The phenolic compounds and some of the acids are degradation products from lignin, while the resin acids, fatty acids, terpenes and sterols are residues of extractives presents in the raw material.

ECF and TCF Wastewaters Treatability

The biological treatment of the effluents from ECF and TCF is almost the same. There is a slight difference in the organic matter constitution among these bleaching effluents, but it is less than other parameters such as raw materials, effluents from the unbleached line, than the bleaching effluent itself.

TCF eucalyptus pulp produced an effluent with 3.5 times the BOD and twice the COD than ECF eucalyptus pulp effluent. Similarly, TCF bleaching effluent had approximately twice the COD in softwood than the ECF effluents. The larger amounts of COD and TOC in the TCF effluents can be explained because the bleaching reagents used in the TCF sequences (O_3, H_2O_2) are less selective towards residual lignin than the ClO_2 use in the ECF sequences. Bleaching of pulps with ozone is known to produce aldehyde and keto groups on carbohydrates, which are highly susceptible to oxidative degradation under alkaline conditions. An alkaline peroxide stage is used to further bleach ozone-treated pulps, resulting in an oxidative degradation of these carbohydrates and thus contributing to higher COD and TOC values in the TCF effluents. Moreover, the hardwood TCF effluents contained more carbohydrates (mainly xylan) than the ECF effluents. An explanation of these differences was that the process conditions in P-stage (long retention time under alkaline conditions) may favour dissolution of xylan from the pulp.

However, while TCF effluent contains more dissolved organic matter, it is less coloured than ECF effluent, mainly because of the action of residual reagents (i.e. H_2O_2) in the TCF effluent. Normal values of colour at 525 nm in TCF effluents are 300 and 1300 C.U. in ECF effluents.

Kraft Pulping: Wastewater Treatment

The typical pulp mill wastewater treatment should include primary treatment (neutralization, screening or sedimentation), principally to remove suspended solids, and biological/secondary treatment. The secondary treatment is mainly done to diminish the organic matter, which is removed by biological degradation, and is particularly useful for the removal of low molecular mass organic matter with a molecular weight of 800 Da or less. Some mills have tertiary treatment to further reduce toxicity, suspended solids, organics or colour.

Secondary biological treatment is applied in most types of pulp and paper mills. The most usual methods are activated sludge and aerated lagoons. Some variations of these systems include the use of filters and sequences reactors—Mobil Bed Bioreactor (MBBR) and Membrane Bioreactors (MBR). Sometimes anaerobic treatment is used followed by an aerobic biological stage.

Aerated ponds and activated sludge methods are the most common treatment systems in pulp and paper industry. In an aerated pond, wastewater is treated through a combination of physical, biological and chemical processes. They have large residence times between 3 and 20 days, and consequently a large volume. They work with low microorganism concentration (low solids concentration) about 100–300 mg/L. These ponds use aeration devices to add oxygen to the wastewater (normally surface turbine aerators or bottom aerators) and mix the contents of the pond, thereby enhancing the microbial activity. However, due to low efficiency levels and the large surface required, the use of aerated lagoons has drastically diminished.

The largest secondary treatment system is activated sludge (60–75% of all the biological effluent treatment plants in pulp and paper industry use activated sludge systems); even in new plants. The advantages of the aerated activated sludge systems compared to the aerobic ponds are that they achieve high removal efficiencies, the process can be well controlled, requires less surface and the microorganisms are adapted to the receiving wastewater. The disadvantages are the high construction and operation costs (especially the energy cost of the aeration systems), the high rate of sludge production and the loss of efficiency due to bulking problems, and consequently, the need to add nutrients to avoid this problem. Sludge handling and nutrient dosage are additional to the energy cost, which is the major component contributing to the operational cost of the biological treatment of process effluents within the pulp and paper industry.

Characteristics of Activated Sludge Treatment

Two main units of the activated sludge plant are the aeration basin and the sedimentation basin. In the aeration basin, the effluent is treated with a culture of microorganisms (the activated sludge), which is present in a high concentration. Figure shows a diagram of a pulp mill treatment with the activated sludge system. Activated sludge plants at kraft pulp mills have a retention time of about 15–48 h. The solids concentration in the activated sludge systems is typically 2000–6000 mg/L. The hydraulic residence time is 4–8 h for a conventional system and the cellular residence time (sludge age) is normally 5–15 days. Normal loads are between 0.05 and 0.1 kg BOD/kg sludge for extended aeration and 0.1–0.3 kg BOD/kg sludge for low load process. The common operating temperature is about 35–37°C and the dissolved oxygen (DO) concentration is 1.5–2.0 ppm. The nutrients concentration in relation to the organic matter is important in effluent treatment. Effluents from the wood processing industry generally have a BOD:N:P ratio of 100:(1–2):(0.15–0.3) and the addition of supplemental nutrients is normally required.

The removal efficiencies reached vary according to the wastewater residence time and the operating conditions. Normal efficiencies figures are between 85 and 98% BOD5 removal and 60–85% for COD removal. For AOX, the reduction is about 40–65%, 40–85% for phosphorus and 20–50% for nitrogen. The overall efficiency of TSS removal using primary and secondary treatment is about 85–90%.

Aerobic Treatability of the Different Effluent Fractions

The COD of treated effluent represents how effective a treatment technology is in its ability to remove the total organic material present in the influent. BOD measurements by themselves

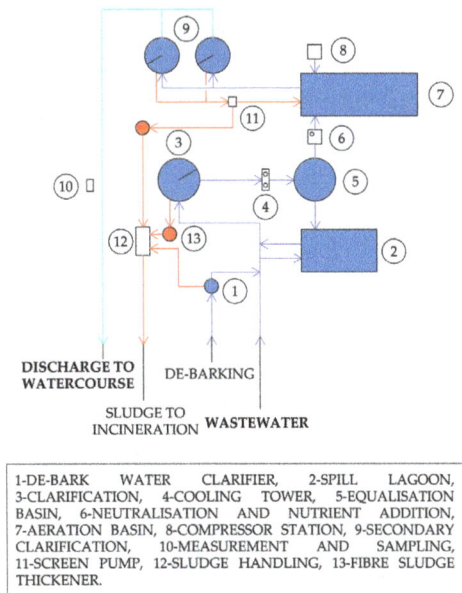

1-DE-BARK WATER CLARIFIER, 2-SPILL LAGOON, 3-CLARIFICATION, 4-COOLING TOWER, 5-EQUALISATION BASIN, 6-NEUTRALISATION AND NUTRIENT ADDITION, 7-AERATION BASIN, 8-COMPRESSOR STATION, 9-SECONDARY CLARIFICATION, 10-MEASUREMENT AND SAMPLING, 11-SCREEN PUMP, 12-SLUDGE HANDLING, 13-FIBRE SLUDGE THICKENER.

Figure: Diagram of a pulp mill treatment plant with activated sludge as biological treatment.

do not quantify the non-biodegradable or slowly biodegradable organic portion of the effluent. Moreover, studies seem to indicate that the residual colour in pulp mill effluents could be linked to the recalcitrant COD.

Recalcitrant organic matter is supposed to be partly responsible for long-term toxicity in receiving waters. As discussed earlier, it is widely reported that the residual recalcitrant organic matter is composed predominantly by high molecular weight components, which are not metabolized due to its size. However, the contributions of high and low molecular weight fractions in bio-treated effluents are dissimilar. In the LMW fraction, a large-scale removal of the chlorinated phenolic compounds, chlorinated resin acids and sterols occurs. In the HMW fraction, the carbohydrates are strongly affected; however, other compounds such as oxidized lignin were less affected.

Some findings are possible by comparing the high molecular weight (HMW) and low molecular weight (LMW) fractions of the acidic and alkaline filtrates post biological treatment. In the alkaline filtrate, the COD and TOC in the HMW fraction increased after treatment. The same behaviour was observed with the AOX and lignin content in the acidic filtrate. This is attributable to the formation of soluble bacterial products or to the adsorption of the LMW into HMW matter. In the LMW filtrates, the COD/TOC decreased after biological treatment, as a result of the large removal of highly oxidized organic carbon. The colour increased in the HMW fractions of acid and alkaline filtrates. The biological treatment often leads to increased colour in ECF bleaching effluents due to the creation of new chromophores in the HMW fractions.

Bulking Problems in the Activated Sludge Systems

Two critical operational aspects of an activated sludge plant are maintaining proper control of the dissolved oxygen (DO) concentration in the aeration tank and preserving a good settling sludge. Reduced settleability results in poor plant performance, as it is difficult to maintain a low concentration of suspended solids in the plant effluent. Activated sludge plants that treat pulp and paper mill wastewaters seem to be particularly prone to this. There are several reasons for poor separation properties, such as filamentous bulking sludge, bulking due to excessive extracellular polymeric substances (EPS), production or formation of small flocs and dispersed biomass. In pulp mill wastewater, bulking is often due to the presence of filamentous bacteria. Common conditions that favour bulking are working at feeding loads ratios out of normal range, deficiencies in nitrogen and phosphorous species or in the level of DO. In kinetic terms, the floc forming microorganisms have a competitive advantage at lower substrate concentrations because that allows the compounds to utilize oxygen and nutrients more efficiently than the not floc forming microorganisms.

The presence of filamentous bacteria was examined for two years in 15 French pulp, paper and board mills wastewater. The study of 25 bulking cases attributed the source in 10 cases to be COD hydraulic overloads, in 8 cases to deficient aeration and in 5 cases to nutrient deficiency.

Partial Closure in Water Circuits

The current market and environmental demands facing pulp and paper mills are the increased closure of the plant circuits and a further reduction or elimination of the wastes produced. The concept of a closed loop mill aims to eliminate discharges to the aquatic environment, recycle and reuse all possible solid and liquid process wastes, and reduce air emissions to the lowest possible quantity and toxicity. However, until today, no kraft mills are operating with complete closure and complete reutilization of the effluents. The most important problem experienced in mills that try to operate for long periods with zero discharge was corrosion caused by chlorides in a number of positions. Nevertheless, great progress has been made in minimizing impacts associated with pulp mill effluents. Water circulation closure methods include dry debarking, effective liquor spilling control, closed screening and washing, condensate stripping and other methods to minimize the loss of wood-derived organic mat ter. Extended and oxygen delignification can significantly reduce bleach plant effluent loads from kraft pulp mills. The bleach plant is the most important source of effluent within a pulp mill and the chlorinated effluents are more complicated to reutilize within the mill. For this reason, an important trend in bleaching development is to reduce volumes and decrease the effluent loads, especially of chlorinated compounds.

Up to now, a complete water closed circulation is not available; nevertheless, a partial closure of the water circuits is possible. This can be done segregating the acid and alkaline effluent streams and recirculating the liquids counter currently from the last bleach stage through the sequence to the brown stock washer. The alkaline effluent could be used for washing the pulp in the unbleached part of the process.

The pulp and paper industry has been considered a large consumer of wood, energy and water, and an important contributor of pollutant discharges to the environment (air, water courses and soil). However, the last decades have seen a lot of effort in creating solutions such as generating less pollutant wastewaters and reducing the amount and load of the emissions to the environment. The implementation of several measures like the dry debarking of wood, the introduction of extended cooking and oxygen delignification, the reuse of condensates, improvements in washing efficiency and especially the total substitution of chlorine, has brought a significant reduction in effluent flows and in the chlorinated and organic loads generated within the mill. In addition, the introduction of end-of-pipe secondary, and even tertiary, treatments have reduced large amounts of pollutant loads to the environment. However, the need for tertiary treatment is not yet well proven; while it purifies the effluent, the energy costs are high and even forms sludge.

Effluent characteristics are dependent on the production process and the raw materials. ECF eucalyptus pulp production is increasing appreciably but not much information on its effluents is available. The main difference between softwood and eucalyptus pulps is in the kappa number: the kappa number is mainly formed by lignin content in softwood pulp, and the Hexenuronic acids are important contributors to kappa number in eucalyptus pulp. Hence, the bleaching conditions for eucalyptus are less severe and consequently the effluents characteristics are different. Eucalyptus bleaching effluents have lower COD, AOX and colour content and higher biodegradability than the softwood effluents.

The environmental impact of effluent loads and the appropriate treatment can be determined by studying the chemical composition and molecular weight distribution of the bleaching effluents. The HMW in hardwood bleaching wastewaters constituted an important but not prevailing fraction of the wastewater composition (30–65% of the total). The hardwood HMW fraction is mainly composed of non-aromatic structural compounds.

Aerated activated sludge is the most common treatment system in pulp mills. BOD5 removals of 85–98% and COD removals of 60–85% are normally achieved with these systems. For AOX, the reduction is about 40–65%, 40–85% for phosphorus and 20–50% for nitrogen. Bulking problems are common in these systems mainly due to nitrogen deficiencies and phosphorous concentration or the level of DO.

Nowadays, plants that apply the best available technologies have their emissions controlled and present minimum environmental impact at the receiving waters.

The new developments are in the way to close even more the internal circuits in the plant, to reduce the flow discharged. Membrane technologies and similar technologies may be key in this regard.

Rubber Industry Wastes Treatment

Rubber is produced from natural or synthetic sources. Natural rubber is obtained from the milky white fluid called latex, found in many plants; synthetic rubbers are produced from unsaturated hydrocarbons.

Long before Colombus arrived in the Americas, the native South Americans were using rubber to produce a number of water-resistant products. The Spaniards tried in vain to copy these products (shoes, coats and capes), and it was not until the 18th century that European scientists and manufacturers began to use rubber successfully on a commercial basis. The British inventor and chemist Charles Macintosh, in 1823, established a plant in Glasgow for the manufacture of waterproof cloth and the rainproof garments with which his name has become synonymous.

A major breakthrough came in the mid-19th century with the development of the process of vulcanisation. This process gives increased strength, elasticity, and resistance to changes in temperature. It also renders rubber impermeable to gases and resistant to heat, electricity, chemical action and abrasion. Vulcanised rubber also exhibits frictional properties highly desired for pneumatic tyre application.

Crude latex rubber has few uses. The major uses for vulcanised rubber are for vehicle tyres and conveyor belts, shock absorbers and anti-vibration mountings, pipes and hoses. It also serves some other specialist applications such as in pump housings and pipes for handling of abrasive sludges, power transmission belting, diving gear, water lubricated bearings, etc.

Natural rubber is extracted from rubber producing plants, most notably the tree Hevea brasiliensis, which originates from South America. Nowadays, more than 90% of all natural rubber comes from these trees in the rubber plantations of Indonesia, the Malay Peninsula and Sri Lanka. The common name for this type of rubber is Para rubber.

The rubber is extracted from the trees in the form of latex. The tree is 'tapped'; that is, a diagonal incision is made in the bark of the tree and as the latex exudes from the cut it is collected in a small cup. The average annual yield is approximately 2 ½ kg per tree or 450kg per hectare, although special high-yield trees can yield as much as 3000kg per hectare each year.

The gathered latex is strained, diluted with water, and treated with acid to cause the suspended rubber particles within the latex to coagulate. After being pressed between rollers to form thin sheets, the rubber is air (or smoke) dried and is then ready for shipment.

Synthetic Rubber

There are several synthetic rubbers in production. These are produced in a similar way to plastics, by a chemical process known as polymerisation. They include neoprene, Buna rubbers, and butyl rubber. Synthetic rubbers have usually been developed with specific properties for specialist applications. The synthetic rubbers commonly used for tyre manufacture are styrene-butadiene rubber and butadiene rubber (both members of the Buna family). Butyl rubber, since it is gas-impermeable, is commonly used for inner tubes. Table below shows typical applications of various types of rubber.

Type of rubber	Application
Natural rubber	Commercial vehicles such as lorries, buses and trailers.
Styrene-butadiene rubber (SBR) and Butadiene rubber (BR)	Small lorries, private cars, motorbikes and bicycles
Butyl rubber (IIR)	Inner tubes.

Table: Applications of different classes of rubber in the manufacture of vehicle tyres.

The raw materials that make up tyres are natural and synthetic rubbers, carbon, nylon or polyester cord, sulphur, resins and oil. During the tyre making process, these are virtually vulcanised into one compound that is not easily broken down.

Production of Rubber Products

The modern process of rubber manufacture involves a sophisticated series of processes such as mastication, mixing, shaping, moulding and vulcanisation. Various additives are included during the mixing process to give desired characteristics to the finished product. They include:

Polymers	Vulcanisation accelerators
Activators	Vulcanisation agents
Fillers (carbon black)	Fire retardants
Anti-degradants	Colorants or pigments
Plasticisers	Softeners

Fillers are used to stiffen or strengthen rubber. Carbon black is an anti-abrasive and is commonly used in tyre production. Pigments include zinc oxide, lithopone, and a number of organic dyes. Softeners, which are necessary when the mix is too stiff for proper incorporation of the various ingredients, usually consist of petroleum products, such as oils or waxes; pine tar; or fatty acids. The moulding of the compound is carried out once the desired mix has been achieved and vulcanisation is often carried out on the moulded product.

Vulcanisation

To understand the process of vulcanisation it is worth discussing, briefly, the molecular structure of rubber. Crude latex is made up of a large number of very long, flexible, molecular chains. If these chains are linked together to prevent the molecules moving apart, then the rubber takes on its characteristic elastic quality. This linking process is carried out by heating the latex with sulphur (other vulcanising agents such as selenium and tellurium are occasionally used but sulphur is the most common). There are two common vulcanising processes.

- Pressure vulcanisation. This process involves heating the rubber with sulphur under pressure at a temperature of 1500 C. Many articles are vulcanised in moulds that are compressed by a hydraulic press.

- Free vulcanisation. Used where pressure vulcanisation is not possible, such as with continuous, extruded products, it is carried out by applying steam or hot air. Certain types of garden hose, for example, are coated with lead, and are vulcanised by passing high-pressure steam through the opening in the hose.

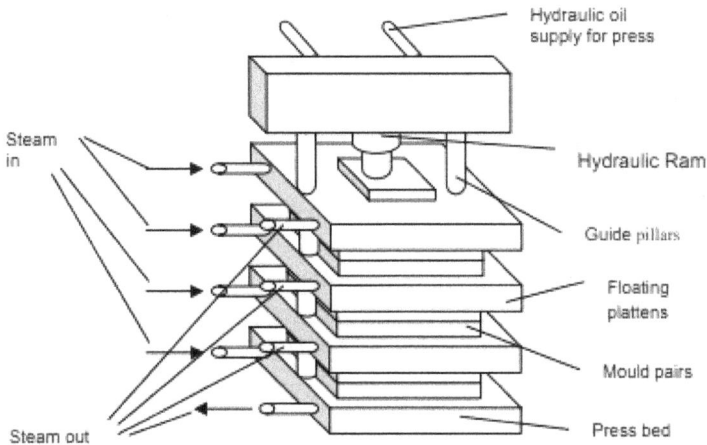

The proportion of natural and synthetic rubber used for tyre manufacture depends on the application of the particular tyre.

Truck tyre tread (in %)		Passenger vehicle tyre tread (in %)
Mineral oil	13	20 −24
Carbon black	30	33 − 37
Rubber – of which	57	40 − 45
Natural rubber	65	25
BR & SBR	35	75

Table: Composition of typical tyre tread for commercial and passenger vehicles. (TOOL 1996)

Reclaim or Recycle Rubber

Rubber recovery can be a difficult process. There are many reasons, however why rubber should be reclaimed or recovered:

- Recovered rubber can cost half that of natural or synthetic rubber.

- Recovered rubber has some properties that are better than those of virgin rubber.

- Producing rubber from reclaim requires less energy in the total production process than does virgin material.

- It is an excellent way to dispose of unwanted rubber products, which is often difficult.

- It conserves non-renewable petroleum products, which are used to produce synthetic rubbers.

- Recycling activities can generate work in developing countries.

- Many useful products are derived from reused tyres and other rubber products.

- If tyres are incinerated to reclaim embodied energy then they can yield substantial quantities of useful power. In Australia, some cement factories use waste tyres as a fuel source.

Tyre Reuse and Recovery in Developing Countries

There is an enormous potential for reclamation and reuse of rubber in developing countries. There is a large wastage of rubber tyres in many countries and the aim of this brief is to give some ideas for what can be done with this valuable resource. Whether rubber tyres are 2reused, reprocessed or hand crafted into new products, the end result is that there is less waste and less environmental degradation as a result.

In developing countries, there is a culture of reuse and recycling. Waste collectors roam residential areas in large towns and cities in search of reusable articles. Some of the products that result from the reprocessing of waste are particularly impressive and the levels of skill and ingenuity are high. Recycling artisans have integrated themselves into the traditional market place and have created a viable livelihood for themselves in this sector. The process of tyre collection and reuse is a task carried out primarily by the informal sector. Tyres are seen as being too valuable to enter the waste stream and are collected and put to use.

In Karachi, Pakistan, for example, tyres are collected and cut into parts to obtain secondary materials which can be put to good use. The beads of the tyres are removed and the rubber removed by burning to expose the steel. The tread and sidewalls are separated – the tread is cut into thin strips and used to cover the wheels of donkey carts, while the sidewalls are used for the production of items such as shoe soles, slippers or washers.

Figure: Manual Separation of the Tread from the Sidewalls, Karachi, Pakistan

Recovery Alternatives

There are many ways in which tyres and inner tubes can be reused or reclaimed. The waste management hierarchy dictates that re-use, recycling and energy recovery, in that order, are superior to disposal and waste management options. This hierarchy is outlined in Table below.

Kind of recovery		Recovery process
Product reuse	Repair	• Retreading • Regrooving
	Physical reuse	• Use as weight • Use of form • Use of properties • • Use of volume
Material reuse	Physical	• Tearing apart • Cutting • Processing to crumb
	Chemical	• Reclamation
	Thermal	• Pyrolysis • Combustion
Energy reuse		• Incineration

Table: Principal rubber recycling processing paths (adapted from van Baarle)

Product re-use

Damaged tyres are, more often than not, repaired. Tubes can be patched and tyres can be repaired by one of a number of methods. Regrooving is a practice carried out in many developing countries where regulations are slacker and standards are lower (and speeds are lower) than in the West. It is often carried out by hand and is labour intensive.

The use of retread tyres saves valuable energy and resources. A new tyre requires 23L of crude oil equivalent for raw materials and 9L for process energy compared with 7L and 2L respectively for retreading. Tyres of passenger vehicles can generally be retreaded only once while truck and bus tyres can be retreaded up to six times. Retreading is a well-established and acceptable (in safety terms) practice. The process involves the removal of the remaining tread and the application and vulcanisation of a new tread (the 'camel back') onto the remaining carcass. In Nairobi about 10,000 tyres a week are received for retreading (Ahmed).

Secondary reuse of whole tyres is the next step in the waste management hierarchy. Tyres are often put to use because of their shape, weight, form or volume. Some examples of secondary use in industrialised countries include use for erosion control, as tree guards, in artificial reefs, fences or as garden decoration. In developing countries wells can be lined with old tyres, docks are often lined with old tyres which act as shock absorbers, and similarly crash barriers can be constructed from old tyres. Old inner tubes also have many uses; swimming aids and water containers being two simple examples.

Figure: Following the grooves is a Labour –intensive process.

Material Re-use

The next step in our hierarchy involves the material being broken down and reused for the production of a new product. As mentioned earlier, in developing countries this hand reprocessing of rubber products to produce consumer goods is well established and the variety of products being made from reclaimed tyres and tubes is astonishing. The rubber used in tyres is a relatively easy material to reform by hand. It behaves in a similar manner to leather and has in fact replaced leather for a number of applications.

The tools required for making products directly from tyre rubber are not expensive and are few in number. Shears, knives, tongs, hammers, etc., all common tools found in the recyclers' workshop, along with a wide range of improvised tools for specialised applications. Shoes, sandals, buckets, motor vehicle parts, doormats, water containers, pots, plant pots dustbins and bicycles pedals are among the products manufactured.

Another way in which physical reuse can be achieved is by reducing the tyre to a granular form and then reprocessing. This can be a costly process and there has to be a manufacturer willing to purchase the granules. Crumb rubber from the retreading process can be used in this way, as it is a good quality granulated rubber. The reprocessing techniques used are similar to those described in earlier chapters. Granulate tends to be used for low-grade products such as automobile floor mats, shoe soles, rubber wheels for carts and barrows, etc., and can be added to asphalt for road construction, where it improved the properties of this material.

Figure: Garbage containers made from Truck tyres. Manila, The Philippines

Chemical and Thermal Recovery

This type of recovery is not only lower in the waste management hierarchy, but is also a higher technology requiring sophisticated equipment. The applicability of such technologies for small-scale applications in developing countries is very limited. We will therefore look only very briefly at a couple of processes. Chemical recovery is the process of heating waste rubber reclaim, treating it with chemicals and then processing the rubber mechanically.

- Acid reclamation – uses hot sulphuric acid to destroy the fabric incorporated in the tyre and heat treatment to render the scrap rubber sufficiently plastic to allow its use as a filler with batches of crude rubber.

- Alkali recovery - Reclaimed rubber, treated by heating with alkali for 12 to 30 hours, can be used as an adulterant of crude rubber to lower the price of the finished article. The amounts of reclaimed rubber that are used depend on the quality of the article to be manufactured.

One form of thermal recovery is pyrolysis. This involves heating the tyre waste in the absence of oxygen which causes decomposition into gases and constituent parts. It is a technology which is still immature in the tyre-reprocessing field.

Energy Recovery

Tyres consist of around 60% hydrocarbons, which is a store of energy that can be recovered by incineration. The heat produced can be used directly in processes such as cement making, or to raise steam for a variety of uses, including electricity generation. Again, this technology requires sophisticated plant and its application is limited when looking at small-scale enterprise.

Landfill

Landfill is the final step in the waste management hierarchy. The landfill disposal of tyres, if properly managed, does not constitute an environmental problem. However, concerns about conserving resources and energy have seen an increasing opposition to landfilling. Also, public sanitation and municipal waste management is often ineffective in developing countries and scrap tyres are often found littering the streets.

Timber Industry Wastes Treatment

Treated timber contains chemicals that can be harmful to the environment, your health or your children's health. It should be safely disposed of, and never burnt or buried.

Treated timber offcuts

- Only put small amounts of treated timber offcuts in your regular rubbish bin.

- Dispose of treated timber from larger household building or demolition jobs at a licenced landfill site.

- Dispose of treated timber sawdust correctly: double-bag large amounts and take them to a licensed landfill site.

You can leave treated timber out for council pick-ups, but try to put it out early on pick up day to reduce the chance of people taking it.

Do not

- use treated timber sawdust or wood shavings for mulch, compost, or animal bedding.

- use leftover timber and offcuts to make animal housing.

- mix treated timber with other timber products for recycling – most treated timber cannot be recycled.

- leave CCA-treated timber where others may take it and use it for firewood – for example, on your nature strip.

Burning treated timber releases toxic chemicals.

- Never burn treated timber or treated timber waste in outdoor fires, stoves, fireplaces or in confined spaces.

- Never use it to cook food.

Treated Timber Ash

Ash from copper chrome arsenic (CCA) treated timber is toxic and may contain more than 10% of its weight as heavy metal residue, including arsenic.

Large amounts of toxic ash can be a big problem when properties are damaged by bushfires or other fires.

- Wear personal protective equipment (PPE) to clean up the ash, double-bag it and take it to a licensed landfill site.

- Do not bury it.

Soft Drink Waste Treatment

The history of carbonated soft drinks dates back to the late 1700s, when seltzer, soda, and other waters were first commercially produced. The early carbonated drinks were believed to be effective against certain illnesses such as putrid fevers, dysentery, and bilious vomiting. In particular, quinine tonic water was used in the 1850s to protect British forces abroad from malaria.

The biggest breakthrough was with Coca-Cola, which was shipped to American forces wherever they were posted during World War II. The habit of drinking Coca-Cola stayed with them even after they returned home. Ingredients for the beverage included coca extracted from the leaves of the Bolivian Coca shrub and cola from the nuts and leaves of the African cola tree. The first Coca-Cola drink was concocted in 1886. Since then, the soft drink industry has seen its significant growth.

Table lists the top 10 countries by market size for carbonated drinks, with the United States leading the pack with the largest market share. In 1988 the average American's consumption of soft drinks was 174 L/year; this figure has increased to approximately 200 L/year in recent years. In 2001, the retail sales of soft drinks in the United States totalled over $61 billion. The US soft drink industry features nearly 450 different products, employs more than 183,000 nationwide and pays more than $18 billion annually in state and local taxes.

Table Top Ten World Market Size in Carbonated Soft Drinks, 1988

Rank	Country	1000 million liters
1	United States	42.7
2	Mexico	8.4
3	China	7.0
4	Brazil	5.1
5	West Germany	4.6
6	United Kingdom	3.5
7	Italy	2.6
8	Japan	2.5
9	Canada	2.4
10	Spain	2.3

The soft drink industry uses more than 12 billion gallons of water during production every year. Therefore, the treatment technologies for the wastewater resulting from the manufacturing process cannot be discounted.

Composition of Soft Drinks

The ingredients of soft drinks can vary widely, due to different consumer tastes and preferences. Major components include primarily water, followed by carbon dioxide, caffeine, sweeteners, acids, aromatic substances, and many other substances present in much smaller amounts. Calories and components of major types of soft drinks.

Water

The main component of soft drinks is water. Regular soft drinks contain 90% water, while diet soft drinks contain up to 99% water. The requirement for water in soft drink manufacturing is that it must be pure and tasteless. For this reason, some form of pre-treatment is required if the tap water used has any kind of taste. The pre-treatment can include coagulation–flocculation, filtration, ion exchange, and adsorption.

Carbon Dioxide

The gas present in soft drinks is carbon dioxide. It is a colourless gas with a slightly pungent odor. When carbon dioxide dissolves in water, it imparts an acidic and biting taste, which gives the drink a refreshing quality by stimulating the mouth's mucous membranes. Carbon dioxide is delivered to soft drink factories in liquid form and stored in high-pressure metal cylinders.

Carbonation can be defined as the impregnation of a liquid with carbon dioxide gas. When applied to soft drinks, carbonation makes the drinks sparkle and foam as they are dispensed and consumed. The escape of the carbon dioxide gas during consumption also enhances the aroma since the carbon dioxide bubbles drag the aromatic components as they move up to the surface of the soft drinks. The amount of the carbon dioxide gas producing the carbonation effects is specified in volumes, which is defined as the total volume of gas in the liquid divided by the volume of the liquid. Carbonation levels usually vary from one to a few known drinks.

In addition, the presence of carbon dioxide in water inhibits microbiological growth. It has been reported that many bacteria die in a shorter time period in carbonated water than in noncarbonated water.

Caffeine

Table 7.2 List of Energy and Chemical Content per Fluid Ounce

Flavor types	Calories	Carbohydrates (g)	Total sugars (g)	Sodium (mg)	Potassium (mg)	Phosphorus (mg)	Caffeine (mg)	Aspartame (mg)
Regular								
Cola or Pepper	12–14	3.1–3.6	3.1–3.6	0–2.3	0–1.5	3.3–6.2	2.5–4.0	0
Caffeine-free cola or Pepper	12–15	3.1–3.7	3.1–3.7	0–2.3	0–1.5	3.3–6.2	0	0
Cherry cola	12–15	3.0–3.7	3.0–3.7	0–1.2	0–1.0	3.9–4.5	1.0–3.8	0
Lemon-lime (clear)	12–14	3.0–3.5	3.0–3.5	0–4.6	0–0.3	0–0.1	0	0
Orange	14–17	3.4–4.3	3.4–4.3	1.1–3.5	0–1.4	0–5.0	0	0
Other citrus	10–16	2.5–4.1	2.5–4.1	0.8–4.1	0–10.0	0–0.1	0–5.3	0
Root beer	12–16	3.1–4.1	3.1–4.1	0.3–5.1	0–1.6	0–1.6	0	0
Ginger ale	10–13	2.6–3.2	2.6–3.2	0–2.3	0–0.3	0–trace	0	0
Tonic water	10–12	2.6–2.9	2.6–2.9	0–0.8	0–0.3	0–trace	0	0
Other regular	12–18	3.0–4.5	3.0–4.5	0–3.5	0–2.0	0–7.8	0–3.6	0
Juice added	12–17	3.0–4.2	3.0–4.2	0–1.8	2.5–10.0	0–6.2	0	0
Diet								
Diet cola or pepper	<.1	0–0.1	0	0–5.2	0–5.0	2.1–4.7	0–4.9	0–16.0
Caffeine-free diet cola, pepper	<.1	0–0.1	0	0–6.0	0–10.0	2.1–4.7	0	0–16.0
Diet cherry cola	<.1	0–<0.04	0–trace	0–0.6	1.5–5.0	2.3–3.4	0–3.8	15.0–15.6
Diet lemon-lime	<.1	0–0.1	0	0–7.9	0–6.9	0–trace	0	0–16.0
Diet root beer	<.2	0–0.4	0	3.3–8.5	0–3.0	0–1.6	0	0–17.5
Other diets	<.6	0–1.5	0–1.5	0–8.0	0.3–10.1	0–trace	0–5.8	0–17.0
Club soda, Seltzer, sparkling water	0	0	0	0–8.1	0–0.5	0–0.1	0	0
Diet juice added	<.3	0.1–0.5	0.1–0.5	0–1.8	0–9.0	0–5.0	0	11.4–16.0

Caffeine is a natural aromatic substance that can be extracted from more than 60 different plants including cacao beans, tea leaves, coffee beans, and kola nuts. Caffeine has a classic bitter taste that enhances other flavours and is used in small quantities.

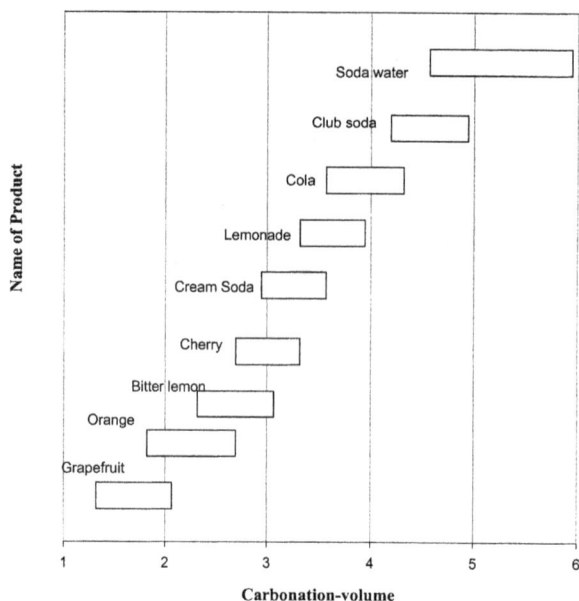

Figure: Carbonation levels of various popular soft drinks.

Sweeteners

Nondiet and diet soft drinks use different types of sweeteners. In nondiet soft drinks, sweeteners such as glucose and fructose are used. Regular (nondiet) soft drinks contain about 7–14% sweeteners, the same as fruit juices such as pineapple and orange. Most nondiet soft drinks are sweetened with high fructose corn syrup, sugar, or a combination of both. Fructose is 50% sweeter than glucose and is used to reduce the number of calories present in soft drinks.

In diet soft drinks, "diet" or "low calorie" sweeteners such as aspartame, saccharin, sucralose, and acesulfame K are approved for use in soft drinks. Many diet soft drinks are sweetened with aspartame, an intense sweetener that provides less than one calorie in a 12 ounce can. Sweeteners remain an active area in food research because of the increasing demand in consumer's tastes and preferences.

Acids

Citric acid, phosphoric acid, and malic acid are the common acids found in soft drinks. The function of introducing acidity into soft drinks is to balance the sweetness and also to act as a preservative. Its importance lies in making the soft drink fresh and thirst-quenching. Citric acid is naturally found in citrus fruits, blackcurrants, strawberries, and raspberries. Malic acid is found in apples, cherries, plums, and peaches.

Other Additives

Other ingredients are used to enhance the taste, color, and shelf-life of soft drinks. These include aromatic substances, colorants, preservatives, antioxidants, emulsifying agents, and stabilizing agents.

The manufacturing and bottling process for soft drinks varies by region and by end products. Generally, the process consists of four main steps: syrup preparation; mixing of carbonic acid, syrup and water; bottling of the soft drink; and inspection.

Syrup Preparation

The purpose of this step is to prepare a concentrated sugar solution. The types of sugar used in the soft drinks industry include beet sugar and glucose. For the production of "light" drinks, sweeteners or a combination of sugar and sweeteners is used instead. After the preliminary quality control, other minor ingredients such as fruit juice, flavourings, extracts, and additives may be added to enhance the desired taste.

Mixing of Carbonic Acid, Syrup, and Water

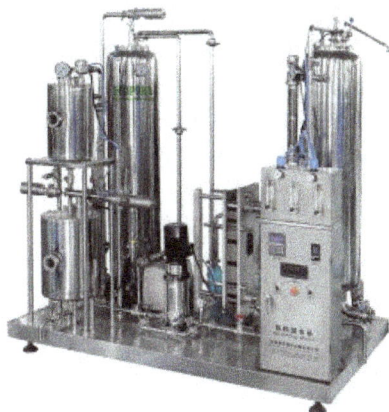

In this second step, the finished syrup, carbonic acid, and water of a fixed composition are mixed together in a computer-controlled blender. This is carried out on a continuous basis. After the completion of the mixing step, the mixed solution is conveyed to the bottling machine via stainless steel piping. A typical schematic diagram of a computer-controlled blender is shown in figure.

Bottling of Soft Drinks

Empty bottles or cans enter the soft drinks factory in palletized crates. A fully automated unpacking machine removes the bottles from the crates and transfers them to a conveyer belt. The unpacking machines remove the caps from the bottles, and then cleaning machines wash the bottles repeatedly until they are thoroughly clean. The cleaned bottles are examined by an inspection machine for any physical damage and residual contamination.

Inspection

This step is required for refillable plastic bottles. A machine that can effectively extract a portion of the air from each plastic bottle is employed to detect the presence of any residual foreign substances. Bottles failing this test are removed from the manufacturing process and destroyed.

A typical bottling machine resembles a carousel-like turret. The speed at which the bottles or cans are filled varies, but generally the filling speed is in excess of tens of thousands per hour. A sealing machine then screws the caps onto the bottles and is checked by a pressure tester machine to see if the bottle or can is properly filled. Finally, the bottles or cans are labeled, positioned into crates, and put on palettes, ready to be shipped out of the factory.

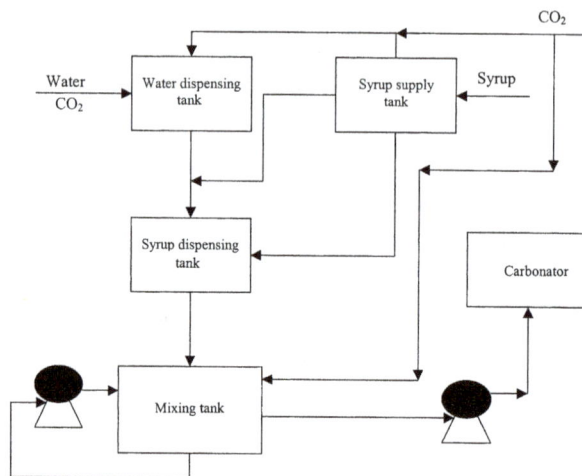

Figure: Schematic diagram of a computer-controlled blender.

Before, during, and after the bottling process, extensive testing is performed on the soft drinks or their components in the laboratories of the bottling plants. After the soft drinks leave the manufacturing factory, they may be subjected to further testing by external authorities.

Characteristics of Soft Drink Wastewater

Soft drink wastewater consists of wasted soft drinks and syrup, water from the washing of bottles and cans, which contains detergents and caustics, and finally lubricants used in the machinery. Therefore, the significant associated wastewater pollutants will include total suspended solids (TSS), 5-day biochemical oxygen demand (BOD5), chemical oxygen demand (COD), nitrates, phosphates, sodium, and potassium gives a list of typical wastewater parameters. As shown, higher organic contents indicate that anaerobic treatment is a feasible process.

Biological Treatment for Soft Drink Wastewater

Biological treatment is the most common method used for treatment of soft drink wastewater because of the latter's organic content. Since BOD5 and COD levels in soft drink wastewaters are moderate, it is generally accepted that anaerobic treatment offers several advantages compared to aerobic alternatives. Anaerobic treatment can reduce BOD5 and COD from a few thousands to a few hundreds mg/L; it is advisable to apply aerobic treatment for further treatment of the wastewater so that the effluent can meet regulations. High-strength wastewater normally has low flow and can be treated using the anaerobic process; low-strength wastewater together with the effluent from the anaerobic treatment can be treated by an aerobic process.

Table: Soft Drink Wastewater Characteristics

Item	Value (mg/L)
COD	1200–8000
BOD5	600–4500
Alkalinity	1000–3500
TSS	0–60
VSS	0–50
NH3-N	150–300
PO4-P	20–40
SO4	7–20
K	20–70
Fe	10–20
Na	1500–2500
Ni	1.2–2.5
Mo	3–8
Zn	1–5
Co	3–8

A complete biological treatment includes optional screening, neutralization/equalization, anaerobic and aerobic treatment or aerobic treatment, sludge separation (e.g., sedimentation or dissolved air flotation), and sludge disposal. Chemical and physical treatment processes (e.g., coagulation and sedimentation/flotation) are occasionally used to reduce the organic content before the wastewater enters the biological treatment process. Since the wastewater has high sugar content, it can promote the growth of filamentous bacteria with lower density. Thus, dissolved air flotation may be used instead of the more commonly used sedimentation.

Aerobic Wastewater Treatment

Owing to the high organic content, soft drink wastewater is normally treated biologically; aerobic treatment is seldom applied. If the waste stream does not have high organic content, aerobic treatment can still be used because of its ease in operation. The removal of BOD and COD can be accomplished in a number of aerobic suspended or attached (fixed film) growth treatment processes. Sufficient contact time between the wastewater and microorganisms as well as certain levels of dissolved oxygen and nutrients are important for achieving good treatment results. An aerobic membrane bioreactor (MBR) for organic removal as well as separation of bio solids can be used in the wastewater treatment.

Aerobic Suspended Growth Treatment Process

Aerobic suspended growth treatment processes include activated sludge processes, sequencing batch reactors (SBR), and aerated lagoons. Owing to the characteristics of the wastewater, the contact time between the organic wastes and the microorganisms must be higher than that for domestic wastewater. Processes with higher hydraulic retention time (HRT) and solids retention time (SRT), such as extended aeration and aerated lagoon, are recommended to be used.

O'Shaughnessy reported that two aerobic lagoons with volume of 267,800 gallons each were used to treat a wastewater from a Coca Cola bottling company. Detention time was 30 days; the design flow was 20,000 gpd. A series of operational problems occurred in the early phase, including a caustic spill incident, continuous clogging of air diffusers, and bad effluent quality due to shock loading (e.g., liquid sugar spill). Failure to meet effluent standards was a serious problem in the treatment plant. It was observed that the effluent $BOD5$ and COD were above 100 and 500 mg/L, respectively. This problem, however, was solved by addition of potassium; the effluent $BOD5$ decreased to 60 mg/L.

Tebai and Hadjivassilis used an aerobic process to treat soft drink wastewater with a daily flow of 560 m^3/day, BOD_5 of 564 mg/L, and TSS of 580 mg/L. Before beginning biological treatment, the wastewater was first treated by physical and chemical treatment processes. The physical treatment included screening and influent equalization; in the chemical treatment, pH adjustment was performed followed by the traditional coagulation/flocculation process. A BOD_5 and COD removal of 43.2 and 52.4%, respectively, was achieved in the physical and chemical treatment processes. In the biological treatment, the BOD_5 loading rate and the sludge loading rate were 1.64 kg BOD5/day m3 and 0.42 kg BOD5/kg MLSS day; the BOD_5 and COD removal efficiencies were 64 and 70%, respectively. The biological treatment was operated at a high-rate mode, which was the main cause for the lower removal efficiencies of BOD_5 and COD.

Attached (Fixed Film) Growth Treatment Processes

Aerobic attached growth treatment processes include a trickling filter and rotating biological contactor (RBC). In the processes, the microorganisms are attached to an inert material and form a biofilm. When air is applied, oxidation of organic wastes occurs, which results in removal of BOD_5 and COD.

In a trickling filter, packing materials include rock, gravel, slag, sand, redwood, and a wide range of plastic and other synthetic materials. Biodegradation of organic waste occurs as it flows over the attached biofilm. Air through air diffusers is provided to the process for proper growth of aerobic microorganisms.

An RBC consists of a series of closely placed circular discs of polystyrene or polyvinyl chloride submerged in wastewater; the discs are rotated through the wastewater. Biodegradation thus can take place during the rotation.

A trickling filter packed with ceramic tiles was used to treat sugar wastewater. The influent BOD5 and COD were 142–203 mg/L and 270–340 mg/L; the organic loading was from 5 to 120 g BOD_5/m^2 day. Removal efficiencies of BOD_5 of 88.5–98% and COD of 67.8–73.6% were achieved. The process was able to cope effectively with organic shock loading up to 200 g COD/L.

An RBC was recommended for treatment of soft drink bottling wastewater in the Cott Corporation. The average wastewater flow rate was 60,000 gpd; its BOD_5 was 3500 mg/L; and TSS was of the order of 100 mg/L. Through a laboratory study and pilot-plant study, it was found that RBC demonstrated the capability of 94% BOD5 removal at average loading rate of 5.3 lb BOD_5 applied per 1000 square feet of media surface.

Anaerobic Wastewater Treatment

The anaerobic process is applicable to both wastewater treatment and sludge digestion. It is an effective biological method that is capable of treating a variety of organic wastes. Because the anaerobic process is not limited by the efficiency of the oxygen transfer in an aerobic process, it is more suitable for treating high organic strength wastewaters (5 g COD/L). Disadvantages of the process include slow start up, longer retention time, undesirable odors from production of hydrogen sulfite and mercaptans, and a high degree of difficulty in operating as compared to aerobic processes. The microbiology of the anaerobic process involves facultative and anaerobic microorganisms, which in the absence of oxygen convert organic materials into mainly gaseous carbon dioxide and methane.

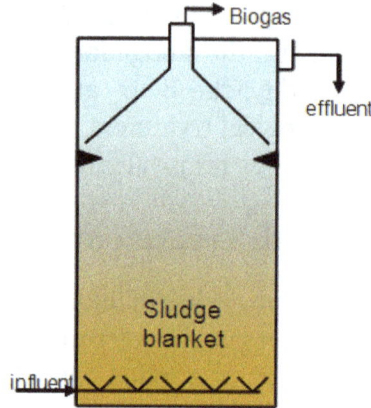

Two distinct stages of acid fermentation and methane formation are involved in anaerobic treatment. The acid fermentation stage is responsible for conversion of complex organic waste (proteins, lipids, carbohydrates) to small soluble product (triglycerides, fatty acids, amino acids, sugars, etc.) by extracellular enzymes of a group of heterogeneous and anaerobic bacteria. These small soluble products are further subjected to fermentation, b-oxidations, and other metabolic processes that lead to the formation of simple organic compounds such as short-chain (volatile) acids and alcohols. There is no BOD5 or COD reduction since this stage merely converts complex organic molecules to simpler molecules, which still exert an oxygen demand. In the second stage (methane formation), short-chain fatty acids are converted to acetate, hydrogen gas, and carbon dioxide in a process known as acetogenesis. This is followed by methanogenesis, in which hydrogen produces methane from acetate and carbon dioxide reduction by several species of strictly anaerobic bacteria.

The facultative and anaerobic bacteria in the acid fermentation stage are tolerant to pH and temperature changes and have a higher growth rate than the methanogenic bacteria from the second stage. The control of pH is critical for the anaerobic process as the rate of methane fermentation remains constant over pH 6.0–8.5. Outside this range, the rate drops drastically. Therefore, maintaining optimal operating conditions is the key to success in the anaerobic process. Sodium bicarbonate and calcium bicarbonate can be added to provide sufficient buffer capacity to maintain pH in the above range; ammonium chloride, ammonium nitrate, potassium phosphate, sodium phosphate, and sodium tripolyphosphate can be added to meet nitrogen and phosphorus requirements.

A number of different bioreactors are used in anaerobic treatment. The microorganisms can be in suspended, attached or immobilized forms. All have their advantages and disadvantages. For example, immobilization is reported to provide a higher growth rate of methanogens since their loss in the effluent can be diminished; however, it could incur additional material costs. Typically, there are three types of anaerobic treatment processes. The first one is anaerobic suspended growth processes, including complete mixed processes, anaerobic contactors, anaerobic sequencing bath reactors; the second is anaerobic sludge blanket processes, including up flow anaerobic sludge blanket (UASB) reactor processes, anaerobic baffled reactor (ABR) processes, anaerobic migrating blanket reactor (AMBR) processes; and the last one is attached growth anaerobic processes with the typical processes of up flow packed bed attached growth reactors, up flow attached growth anaerobic expanded-bed reactors, attached growth anaerobic fluidized-bed reactors, down flow attached growth processes. A few processes are also used, such as covered anaerobic lagoon processes and membrane separation anaerobic treatment process.

It is impossible to describe every system here; therefore, only a select few that are often used in treating soft drink wastewater are discussed in this chapter. Figure shows the schematic diagram of various anaerobic reactors, and the operating conditions of the corresponding reactors are given in Table.

Up Flow Anaerobic Sludge Blanket Reactor

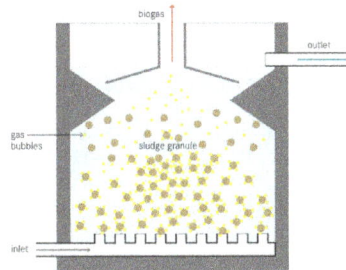

The up flow anaerobic sludge blanket reactor, which was developed by Lettinga, van Velsen, and Hobma in 1979, is most commonly used among anaerobic bioreactors with over 500 installations treating a wide range of industrial wastewaters.

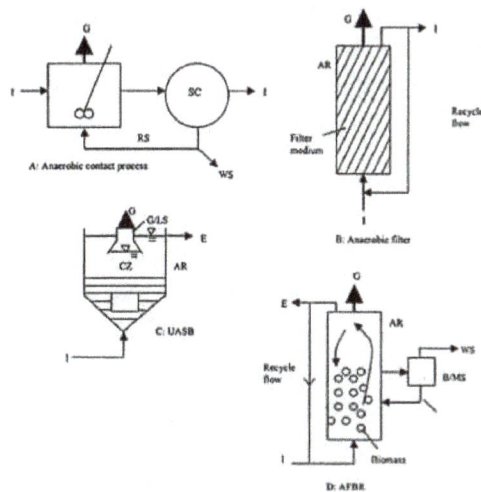

Figure: Schematic diagram of various anaerobic wastewater treatment reactors. AR: anaerobic reactor; B/MS: biofilm/media separator; CZ: clarification zone; E: effluent; G: biogas; G/LS: gas-liquid separator; I: influent; RS: return sludge; SC: secondary clarifier; SZ: sludge zone; WS: waste sludge.

The UASB is essentially a suspended-growth reactor with the fixed biomass process incorporated. Wastewater is directed to the bottom of the reactor where it is in contact with the active anaerobic sludge solids distributed over the sludge blanket. Conversion of organics into methane and carbon dioxide gas takes place in the sludge blanket. The sludge solids concentration in the sludge bed can be as high as 100,000 mg/L. A gas–liquid separator is usually incorporated to separate biogas, sludge, and liquid. The success of UASB is dependent on the ability of the gas–liquid separator to retain sludge solids in the system. Bad effluent quality occurs when the sludge flocs do not form granules or form granules that float.

The UASB can be used solely or as part of the soft drink wastewater treatment process. Soft drink wastewater containing COD of 1.1–30.7 g/L, TSS of 0.8–23.1 g/L, alkalinity of AC, anaerobic contactor; UASB, up flow anaerobic sludge bed; AF, anaerobic filter; AFBR, anaerobic fluidized bed reactor.

Table: Operating Conditions of Common Anaerobic Reactors

Reactor type	AC	UASB	AF	AFBR
Organic loading (kg COD/ m^3-day)	0.48–2.40	4.00–12.01	0.96–4.81	4.81–9.61
COD removal (%)	75–90	75–85	75–85	80–85
HRT (hour)	2–10	4–12	24–48	5–10
Optimal temperature (°C)		30–35 (mesophilic) 49–55 (thermophilic)		
Optimal pH		6.8–7.4		
Optimal total alkalinity (mgCaCO$_3$/L)		2000–3000		
Optimal volatile acids (mg/L as acetic acid)		50–500		

AC, anaerobic contactor; UASB, upflow anaerobic sludge bed; AF, anaerobic filter; AFBR, anaerobic fluidized bed reactor.

1.25–1.93 g CaCO3/L, nitrogen of 0–0.05 gN/L and phosphate of 0.01–0.07 gP/L was treated by a 1.8 L UASB reactor. The pH of wastewater was 4.3–13.0 and temperature was between 20 and 328C. The highest organic loading reported was 16.5 kg COD m23 day21. A treatment efficiency of 82% was achieved.

The "Biothane" reactor is a patented UASB system developed by the Bioethane Corporation in the United States. Its industrial application in wastewater treatment systems was described by Zoutberg and Housley. The wastewater mainly consists of waste sugar solution, product spillage, and wastewater from the production lines. The flow rate averages about 900 m3/day with an average BOD and COD load of 2340 and 3510 kg/day, respectively. The soft drink factory was then producing 650 106 L of product annually, with three canning lines each capable of producing 2000 cans/min and three bottling lines each capable of filling 300 bottles/min. A flow diagram of the "Biothane" wastewater treatment plant is shown in figure. Monitoring of the plant could be performed on or off site. A supervisory control and data acquisition system (SCADA) was responsible for providing continuous monitoring of the process and onsite equipment.

In normal operation, COD removal of 75–85% was reported with 0.35 m3 of biogas produced per kg COD.

Anaerobic Filters

The anaerobic filter was developed by Yong and McCarty in the late 1960s. It is typically operated like a fixed-bed reactor, where growth-supporting media in the anaerobic filter contacts wastewater. Anaerobic microorganisms grow on the supporting media surfaces and void spaces in the media particles. There are two variations of the anaerobic filters: up flow and down flow modes. The media entraps SS present in wastewater coming from either the top (down flow filter) or the bottom (up flow filter). Part of the effluent is recycled and the magnitude of the recycle stream determines whether the reactor is plug-flow or completely mixed. To prevent bed clogging and high head loss problems, backwashing of the filter must be periodically performed to remove biological and inert solids trapped in the media. Turbulent fluid motion that accompanies the rapid rise of the gas bubbles through the reactor can be helpful to remove solids in the media.

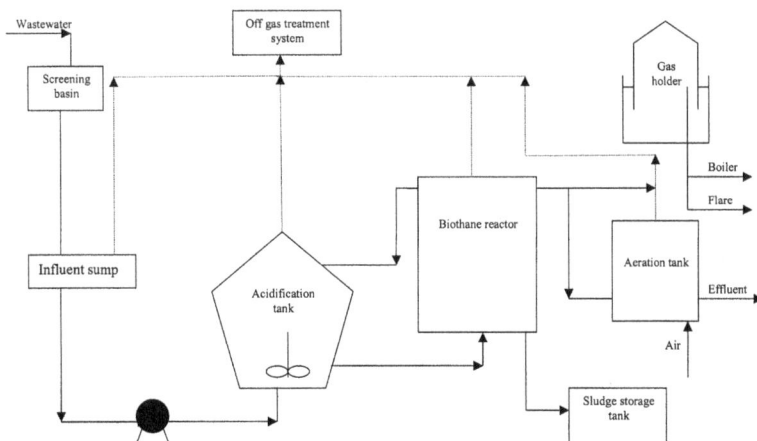

Figure: Flow diagram of the "Biothane" wastewater treatment plant.

Siino used an anaerobic filter to treat soluble carbohydrate waste (soft drink wastewater). At an HRT of 1.7 days, organic loading of 44–210 lb. COD/1000 ft3/day, and SRT of 137 days, removal of 85–90% of COD ranging from 1200 to 6000 mg/L can be achieved. The percentage of methane ranged from 60 to 80%; its product was 0.13–0.68 ft3/day. COD removal efficiency (E %) can be estimated by the following equation:

$$E = 93(1-1:99/HRT)$$

Anaerobic Fluidized Bed Reactor

Soft drink wastewater can also be treated by an anaerobic fluidized bed reactor (AFBR), which is similar in design to the up flow expanded-bed reactor. Influent wastewater enters the reactor from the bottom. Biomass grows as a biolayer around heavy small media particles. At a certain up flow velocity, the weight of the media particles equals the drag force exerted by the wastewater. The particles then become fluidized and the height of the fluidized bed is stabilized.

Packing size of 0.3–0.8 mm and up flow liquid velocities of 10–30 m/hour can be used in order to provide 100% bed expansion. The high flow velocity around the media particles provides good mass transfer of the dissolved organic matter from the bulk liquid to the particle surface. The bed depth normally ranges from 4 to 6 m. Sand, diatomaceous earth, anion and cation exchange resins, and activated carbon can be used as packing materials. The overall density of media particles decreases as the biomass growth accumulates on the surface areas. This can cause the biomass attached media particles to rise in the reactor and eventually wash out together with the effluent. To prevent this from occurring, a portion of the biomass attached particles is wasted and sent to a mechanical device where the biomass is separated from the media particles. The cleaned particles are then returned to the reactor, while the separated biomass is wasted as sludge. Owing to the high turbulence and thin biofilms developed in an AFBR, biomass capture is relatively weak; therefore, an AFBR is better suited for wastewater with mainly soluble COD.

Borja and Banks reported that bentonite, saponite, and polyurethane were respectively used as the suspended support materials for three AFBRs. The composition and parameters of the soft drink wastewater were: total solids (TS) of 3.7 g/L; TSS of 2.9 g/L; volatile suspended solids (VSS) of 2.0 g/L; COD of 4.95 g/L; volatile acidity (acetic acid) of 0.12 g/L; alkalinity of 0.14 g $CaCO_3$/L; ammonium of 5 mg/L; phosphorus of 12 mg/L; pH of 4.8. The average COD removal efficiencies for the three reactors were 89.9% for bentonite, 93.3% for saponite, 91.9% for polyurethane. The amount of biogas produced decreases with increasing HRT. The percentages of methane were 66.0% (bentonite), 72.0% (saponite), and 69.0% (polyurethane).

Borja and Banks used zeolite and sepiolite as packing materials in AFBRs to treat soft drink wastewater. On average, the COD removal of 77.8% and yield coefficient of methane was 0.325 L CH_4/g COD destroyed. The effluent pH was around 7.0–7.3 in all reactors. The content of methane in the biogas ranges from 63 to 70%.

Hickey and Owens conducted a pilot-plant study on the treatment of soft drink bottling wastewater using an AFBR. Diluted soda syrup was used as the substrate, and nitrogen and phosphorus were added with a COD:N: P ratio of 100 : 3 : 0.5. An organic loading rate of 4.0–18.5 kg COD/m³ day results in BOD_5 and COD removal of 61–95% and 66–89%, respectively. Within this organic loading range, the solids production varies from 0.029 to 0.083 kg TSS/kg COD removed. Methane gas was produced at a rate of 0.41 L/g COD destroyed. The composition of the biogas consists of 60% methane and 40% CO_2.

Combined Anaerobic Treatment Process

A combination of different anaerobic reactors has been used to treat soft drink wastewater. It has been reported that treatment efficiency and liability for combined reactors are better than those of a single type of reactor. Several examples are given below.

Stronach reported that a combination of up flow anaerobic sludge blanket reactor, anaerobic fluidized-bed reactor, and anaerobic filter was used to treat fruit processing and soft drink wastewater with TSS, COD, and pH of 160–360 mg/L, 9–15 g/L, and 3.7–6.7, respectively. The organic loadings were 0.75–3.00 kg COD m^{-3} day^{-1} for all three different reactors. COD removal efficiency .79% was achieved. The AFBR performed better than the UASB and the AF in terms of COD removal efficiency and pH stability; however, the methane production was the greatest in the UASB.

Vicenta reported that a 68 L semipilot scale AF installed in series with a UASB was used to treat bottling wastes (bottling washing water and spent syrup wastewater). At an organic loading of 0.59 and 0.88 kg COD m^{-3} day^{-1} for the AF and UASB respectively, an overall COD removal of 75% was achieved. The hydraulic retention time (HRT) for the AF and UASB was maintained at 3.4 and 2.2 days, respectively. An average gas yield of 0.83 L per L of influent was produced.

Silverio used a series of UASB and up flow AF and trickling filter to treat bottling wastewater with pH of 7.6, COD of 7500 mg/L, TSS of 760 mg/L, and alkalinity of 370 mg CaCO$_3$/L, respectively. The total capacity of the reactors in series is 239 L. An organic loading of 2.78 kg COD m^{-3} day^{-1} and HRT of 2.5 days achieved COD removal of 73% and gas yield of 1 L per L of wastewater in the UASB. The COD level of the effluent from the AF after the UASB further dropped to 550 mg/L and corresponded to a removal efficiency of 87%. The HRT and organic loading in the AF were 2.2 days and 0.88 kg COD m^{-3} day^{-1}, respectively. Incorporation of the trickling filter further reduced the COD level of the effluent to 100 mg/L. All biological treatment processes are discussed in detail in Wang and Wang.

Bakery Waste Treatment

Flow diagram of bakery waste water treatment plant

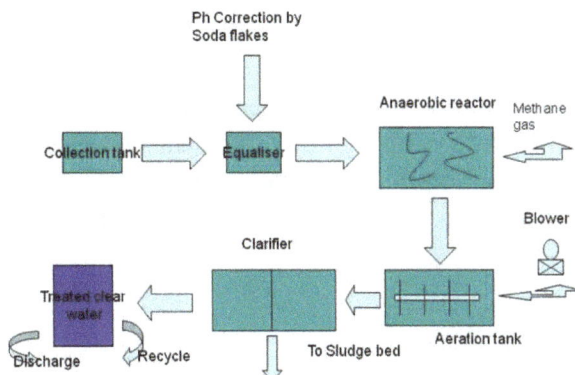

The bakery industry is one of the world's major food industries and varies widely in terms of production scale and process. Traditionally, bakery products may be categorised as bread and bread roll products, pastry products (e.g. pies and pasties) and speciality products (e.g. cake, biscuits, donuts and speciality breads).

The major equipment includes miller, mixer/kneading machine, bun and bread former, fermenter, bake ovens, cold stage and boilers. The main processes are milling, mixing, fermentation, baking and storage. Fermentation and baking are normally operated at 40°C and 160°-260°C, respectively. Depending on logistics and the market, the products can be stored at 4-20°C.

Waste-Water in Bakeries

Waste-water in bakeries is primarily generated from cleaning operations including equipment cleaning and floor washing. It can be characterised as high loading, fluctuating flow and contains rich oil and grease. Flour, sugar, oil, grease and yeast are the major components in the waste.

The ratio of water consumed to products is about 10 in common food industry, much higher than that of 5 in the chemical industry and 2 in the paper and textiles industry. Normally, half of the water is used in the process, while the remainder is used for washing purposes (e.g. of equipment, floor and containers).

Typical values for waste-water production are summarised in tables. Different products can lead to different amounts of waste-water produced. As shown in table, pastry production can result in much more waste-water than the others. The values of each item can vary significantly as demonstrated in table. The waste-water from cake plants has higher strength than that from bread plants. The pH is in acidic to neutral ranges, while the 5-day biochemical oxygen demand (BOD5) is from a few hundred to a few thousand mg/l, which is much higher than that from the domestic wastewater.

The suspended solids (SS) from cake plants is very high. Grease from the bakery industry is generally high, which results from the production operations. The waste strength and flow rate are very much dependent on the operations, the size of the plants and the number of workers.

Generally speaking, in the plants with products of bread, bun and roll, which are termed as dry baking, production equipment (e.g. mixing vats and baking pans) are cleaned dry and floors are swept before washing down. The waste-water from clean-up has low strength and mainly contains flour and grease. On the other hand, cake production generates higher strength waste, which contains grease, sugar, flour, filling ingredients and detergents.

Table: Summary of waste production from the bakery industry

Manufacturer	Products	Waste-water production (litre/ton production)	COD (kg/tonne production)	Contribution to total COD loading (%)
Bread and bread roll	Bread and bread roll	230	1.5	63
Pastry	Pies and sausage rolls	6000	18	29
Speciality	Cake, biscuits, donuts and Persian breads	74	–	–

Due to the nature of the operation, the waste-water strength changes at different operational times. As demonstrated in table, higher BOD5, SS, total solids (TS) and grease are observed from 1 to 3 a.m., which results from lower waste-water flow rate after midnight.

Table: Waste-water characteristics in the bakery industry.

Type of bakery	pH	BOD_5 (mg/l)	SS (mg/l)	TS (mg/l)	Grease (mg/l)
Bread plant	6.9–7.8	155–620	130–150	708	60–68
Cake plant	4.7–8.4	2240–8500	963–5700	4238–5700	400–1200
Variety plant	5.6	1600	1700	–	630
Unspecified	4.7–5.1	1160–8200	650–13430	–	1070–4490

Bakery waste-water lacks nutrients; the low nutrient value gives BOD5:N:P of 284:1:2. This indicates that to obtain better biological treatment results, extra nutrients must be added to the system. The existence of oil and grease also retards the mass transfer of oxygen. The toxicity of excess detergent used in cleaning operations can decrease the biological treatment efficiency. Therefore, the pre-treatment of waste-water is always needed.

Bakery Waste Treatment

Generally, bakery industry waste is nontoxic. It can be divided into liquid waste, solid waste and gaseous waste. In the liquid phase, there are high contents of organic pollutants including chemical oxygen demand (COD), BOD5, as well as fats, oils and greases (FOG) and SS. Waste-water is normally treated by physical, chemical and biological processes.

Pre-Treatment Systems

Pre-treatment or primary treatment is a series of physical and chemical operations, which precondition the waste-water as well as remove some of the wastes. The treatment is normally arranged in the following order: screening, flow equalisation and neutralisation, optional FOG separation, optional acidification, coagulation-sedimentation and dissolved air flotation. The pre-treatment of bakery waste-water is presented in figure.

In the bakery industry, pre-treatment is always required because the waste contains high SS and floatable FOG.

Pre-treatment can reduce the pollutant loading in the subsequent biological and/ or chemical treatment processes; it can also protect process equipment. In addition, pre-treatment is economically preferable in the total process view as compared to biological and chemical treatment.

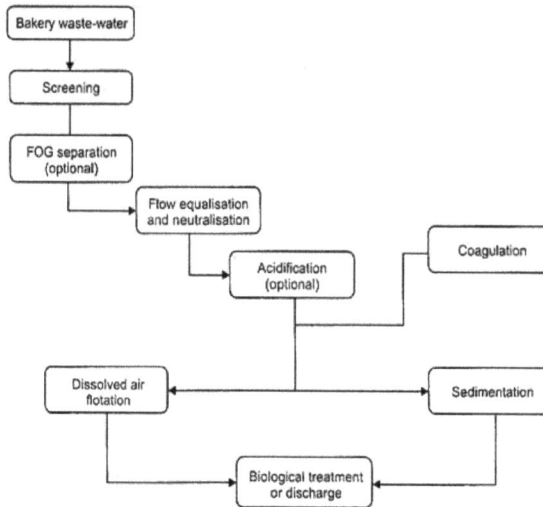

Bakery waste-water pre-treatment system process flow diagram.

Flow Equalisation and Neutralisation

In bakery plants, the waste-water flow rate and loading vary significantly with the time as illustrated in Table. It is usually economical to use a flow equalisation tank to meet the peak discharge demand. However, too long a retention time may result in an anaerobic environment. A decrease in pH and bad odours are common problems during the operations.

Table: Average waste characteristics at specified time interval in a cake plant.

Time interval	pH	BOD_5 (mg/l)	SS (mg/l)	TS (mg/l)	Grease (mg/l)
3 am–8 am	7.9	1480	834	3610	428
9 am–12 am	8.6	2710	1080	5310	457
1 pm–6 pm	8.1	2520	795	4970	486
7 pm–12 pm	8.6	2020	953	3920	739
1 am–3 am	8.9	2520	1170	4520	991

Screening

Screening is used to remove coarse particles in the influent. There are different screen openings ranging from a few μm (termed as microscreen) to more than 100 mm (termed as coarse screen). Coarse screen openings range from 6-150 mm; fine screen openings are less than 6 mm. Smaller opening can have a better removal efficiency; however, operational problems such as clogging and higher head lost are always observed.

Fine screens made of stainless material are often used. The main design parameters include velocity, selection of screen openings and head loss through the screens. Clean operations and waste disposal must be considered. Design capacity of fine screens can be as high as 0.13 m3/sec; the head loss ranges from 0.8-1.4 metre. Depending on the design and operation, BOD5 and SS removal efficiencies are 5-50 per cent and 5-45 per cent, respectively.

FOG Separation

As waste-water may contain high amount of FOG, a FOG separator is thus recommended for installation. The FOG can be separated and recovered for possible reuse, as well as reduce difficulties in the subsequent biological treatment.

Acidification

Acidification is optional, depending on the characteristics of the waste. Owing to the presence of FOG, acid (e.g. concentrated H2SO4) is added into the acidification tank; hydrolysis of organics can occur, which enhances the bio treatability. Grove designed a treatment system using nitric acid to break the grease emulsions followed by an activated sludge process. A BOD5 reduction of 99 per cent and an effluent BOD5 of less than 12 mg/l were obtained at a loading of 40 lb BOD5/1000 ft3 and detention time of 87 hours. The nitric acid also furnished nitrogen for proper nutrient balance for the biodegradation.

Coagulation-Flocculation

Coagulation is used to destabilise the stable fine SS, while flocculation is used to grow the destabilised SS, so that the SS become heavier and larger enough to settle down. The Coagulation-flocculation process can be used to remove fine SS from bakery waste-water. It normally acts as a preconditioning process for sedimentation and/or dissolved air flotation.

The waste-water is preconditioned by coagulants such as alum. The pH and coagulant dosage are important in the treatment results. Liu and Lien reported that 90-100 mg/l of alum and ferric chloride were used to treat waste-water from a bakery that produced bread, cake and other desserts. The wastewater had pH of 4.5, SS of 240 mg/l and COD of 1307 mg/l.

Values of 55 per cent and 95-100 per cent for removal of COD and SS, respectively, were achieved. The optimum pH for removal of SS was 6.0, while that for removal of COD was 6.0-8.0. It was also found that FeCl3 was relatively more effective than alum. Yim used coagulation-flocculation to treat a waste-water with much higher waste strength.

Table: Comparison of different bakery pre-treatment methods.

Coagulant	BOD$_5$		SS		FOG	
	Influent (mg/l)	Removal (%)	Influent (mg/l)	Removal (%)	Influent (mg/l)	Removal (%)
Ferric sulphate	2780	71	2310	94	1450	· 93
Alum	2780	69	2310	97	1450	96

Owing to the higher organic content, SS and FOG, coagulants with high dosage of 1300 mg/l were applied. The optimal pH was 8.0. As shown, removal for the above three items was fairly high, suggesting that the process can also be used for high-strength bakery waste. However, the balance between the cost of chemical dosage and treatment efficiency should be justified.

Sedimentation

Sedimentation, also called clarification, has a working mechanism based on the density difference between SS and the water, allowing SS with larger particle sizes to more easily settle down. Rectangular tanks, circular tanks, combination flocculator-clarifiers and stacked multilevel clarifiers can be used.

Dissolved Air Flotation

Dissolved air flotation (DAF) is usually implemented by pumping compressed air bubbles to remove fine SS and FOG in the bakery waste-water. The waste-water is first stored in an air pressured, closed tank. Through the pressure-reduction valves, it enters the flotation tank. Due to the sudden reduction in pressure, air bubbles form and rise to the surface in the tank. The SS and FOG adhere to the fine air bubbles and are carried upwards. Dosages of coagulant and control of pH are important in the removal of BOD5, COD, FOG and SS.

Other influential factors include the solids content and air/solids ratio. Optimal operation conditions should be determined through the pilot-scale experiments. Liu and Lien used a DAF to treat a waste-water from a large-scale bakery. The waste-water was preconditioned by alum and ferric chloride. With the DAF treatment, 48.6 per cent of COD and 69.8 per cent of SS were removed in 10 minutes at a pressure of 4 kg/cm2 and pH 6.0. Mulligan used DAF as a pre-treatment approach for bakery waste. At operating pressures of 40-60 psi, grease reductions of 90-97 per cent were achieved. The BOD5 and SS removal efficiencies were 33-62 per cent and 59-90 per cent, respectively.

Biological Treatment

The objective of biological treatment is to remove the dissolved and particulate biodegradable components in the waste-water. It is a core part of the secondary biological treatment system. Micro-organisms are used to decompose the organic wastes.

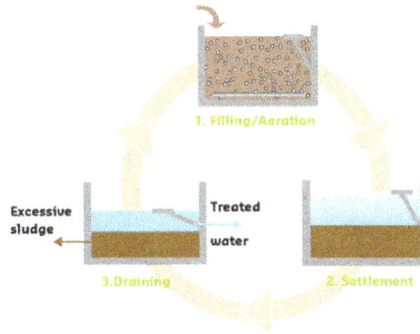

Cyclor - Operating principle

With regard to different growth types, biological systems can be classified as suspended growth or attached growth systems. Biological treatment can also be classified by oxygen utilisation – aerobic, anaerobic and facultative. In an aerobic system, the organic matter is decomposed to carbon dioxide, water and a series of simple compounds. If the system is anaerobic, the final products are carbon dioxide and methane.

Compared to anaerobic treatment, the aerobic biological process has better quality effluent, easier operation, shorter solid retention time, but higher cost for aeration and more excess sludge. When treating high-load influent (COD > 4000 mg/l), the aerobic biological treatment becomes less economic than the anaerobic system. To maintain good system performance, the anaerobic biological system requires more complex operations. In most cases, the anaerobic system is used as a pre-treatment process.

Suspended growth systems (e.g. activated sludge process) and attached growth systems (e.g. trickling filter) are two of the main biological waste-water treatment processes. The activated sludge process is most commonly used in treatment of waste-water. The trickling filter is easy to control and has less excess sludge. It has higher resistance loading and low energy cost. However, high operational cost is its major disadvantage. In addition, it is more sensitive to temperature and has odour problems. Comprehensive considerations must be taken into account when selecting a suitable system.

Aerobic Treatment

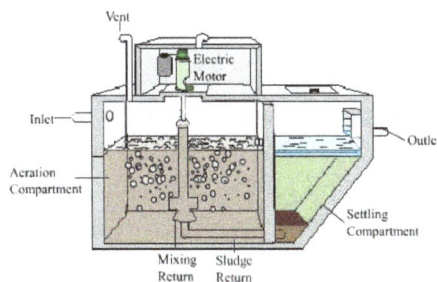

Aerobic Treatment Unit Components

Activated Sludge Process

In the activated sludge process, suspended growth micro-organisms are employed. A typical activated sludge process consists of a pre-treatment process (mainly screening and clarification), aeration tank (bioreactor), final sedimentation and excess sludge treatment (anaerobic treatment and dewatering process).

The final sedimentation separates micro-organisms from the water solution. In order to enhance the performance result, most of the sludge from the sedimentation is recycled back to the aeration tank(s), while the remaining is sent to anaerobic sludge treatment. A recommended complete activated sludge process is given in Fig.

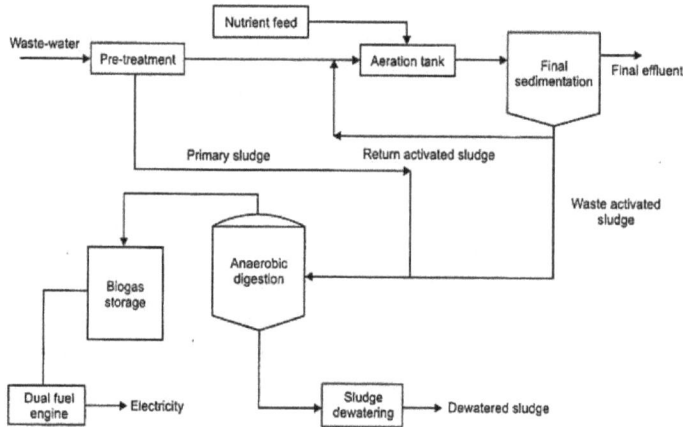

Fig. 27.2. Process flow diagram of activated sludge treatment of bakery waste-water.

The activated sludge process can be a plug-flow reactor (PFR), completely stirred tank reactor (CSTR), or sequencing batch reactor (SBR). For a typical PFR, length-width ratio should be above 10 to ensure the plug flow. The CSTR has higher buffer capacity due to its nature of complete mixing, which is a critical benefit when treating toxic influent from industries. Compared to the CSTR, the PFR needs a smaller volume to gain the same quality of effluent. Most large activated sludge sewage treatment plants use a few CSTRs operated in series. Such configurations can have the advantages of both CSTR and PFR.

The SBR is suitable for treating non-continuous and small-flow waste-water. It can save space, because all five primary steps of fill, react, settle, draw and idle are completed in one tank. Its operation is more complex than the CSTR and PFR; in most cases, auto operation is adopted.

The performance of activated sludge processes is affected by influent characteristics, bioreactor configuration and operational parameters. The influent characteristics are waste-water flow rate, organic concentration (BOD5 and COD), nutrient compositions (nitrogen and phosphorus), FOG, alkalinity, heavy metals, toxins, pH and temperature. Configurations of the bioreactor include PFR, CSTR, SBR, membrane bioreactor (MBR) and so on.

Operational parameters in the treatment are biomass concentration [mixed liquor volatile suspended solids concentration (MLVSS) and volatile suspended solids (VSS)], organic load, food to micro-organisms (F/M), dissolved oxygen (DO), sludge retention time (SRT), hydraulic retention time (HRT), sludge return ratio and surface hydraulic flow load. Among them, SRT and DO are the most important control parameters and can significantly affect the treatment results. A suitable SRT can be achieved by judicious sludge wasting from the final clarifier. The DO in the aeration tank should be maintained at a level slightly above 2 mg/l.

Power Industry Wastes Treatment

Many years have passed since the advent of nuclear power was hailed as providing "electricity too cheap to meter". Nevertheless the main motivation for nuclear power development programmes is to provide an affordable and secure source of electricity both for the short and long term. The cost at which electricity can be provided is therefore a highly important issue, as is the choice of the method for calculating this cost. For many years, the relative costs of different methods of electricity generation have been estimated and compared by a wide range of organizations, including the IAEA, in order to develop a proper perspective.

Since the initial development of nuclear fission reactors, radioactive waste management has often been seen as one of the major problems of nuclear power. Concerns have extended to the costs involved, in particular the cost associated with the disposal of high-level waste or unprocessed spent fuel. This cost has been widely used, not always objectively, by opponents of nuclear power in their arguments. More recently, environmentalist organizations have started to realize that all forms of energy production generate waste and have environmental effects which may be unacceptable, if not adequately controlled. The escalation over the last few years of topics, such as the "greenhouse effect" and "acid rain", into major political issues, has led to more detailed consideration of the waste management aspects from burning fossil fuels. These have hitherto been very loosely regulated, particularly in some parts of the world. We are now at a stage where the management of wastes from nuclear power remains very highly regulated and where the regulations for the control of wastes from fossil fuel power stations are being significantly tightened.

A number of cost studies have recently been completed for different stages of waste management. It was considered useful to collect the results of these studies and to compare them objectively with the waste management costs of electricity production from other energy sources. The comparison can then be used to provide a correct perspective of the economic and environmental aspects of the different means of production of electricity.

The comparison is made for the costs of managing waste generated in the production of electricity from representative nuclear and fossil fuel cases. The associated costs from the third major source of electricity, hydropower, are obviously small and thus not considered here. Both fossil and nuclear fuels can be exploited in a number of different types of plants. Since it would be impractical to consider all possible variants, representative plants having a capacity of 1000 megawatts-electric (MWe) were selected for the assessment, each operating at a capacity factor of 70% for 30 years.

Nuclear and Fossil Fuel Cycles

Fossil fuel cycle: Coal is the leading fossil fuel used for electricity generation in the industrialized world, although the share of gas is increasing rapidly. In some countries, oil is also an important fuel for electricity production, but many try to avoid its use because of possible rapid changes in the price of oil. From the standpoint of waste arisings, oil is somewhere between coal and gas. Because of this, coal and gas have been chosen as the representative fossil cases for the comparison.

Modern coal plants are fired by pulverized coal. Upon combustion. The coal reacts with oxygen to form carbon dioxide (CO_2). The combustion process is accompanied by the production of oxides of nitrogen (NO_x). sulphur dioxide (SO_2), fly ash. and a number of other polluting by-products including radionuclides contained in coal.

Electrical lines in Norway.

For base load electricity production. Two types of gas-fired plants are available. The first is a conventional steam cycle (CSC) plant. More recent units, however, use gas turbines in front of the steam cycle to improve the efficiency of the unit: this combination is called a combined cycle (CC) plant. The combustion process of natural gas is much cleaner than coal. The main combustion products are CO_2 water, and NO_x.

A conventional coal plant and a gas-fuelled combined cycle plant are likely to be the major sources of new fossil-fuelled electricity generation. They represent both ends of the spectrum of challenges associated with waste management from fossil fuels.

Nuclear fuel cycles: Nuclear power plants generate electricity from the heat produced when the nuclei of the atoms of heavy elements are split. The heat is used to produce steam to drive turbines which generate electricity.

Uranium is currently the principal nuclear fuel. It occurs in nature and is mined by conventional mining techniques. It is then processed into a form suitable for use as fuel in a nuclear reactor. Natural uranium contains two main isotopes, uranium-238 and uranium-235. Only the nuclei of the uranium-235 atoms are readily fissile, but uranium-235 accounts for only 0.79r of natural uranium. Some reactors use natural uranium for fuel, but most reactors now use slightly enriched uranium, in which the proportion of uranium-235 atoms has been increased (or enriched) to a few percent. Consequently, most uranium is enriched before it is fabricated into fuel elements for loading into a reactor.

When the spent fuel is removed from the reactor, typically annually, it contains unconsumed uranium, fission products, plutonium. And other heavy elements. It is possible to dissolve the spent fuel and chemically process (reprocess) it in order to extract the unused uranium and plutonium for fuel fabrication and recycling. Alternatively the spent fuel elements can be disposed of directly as waste, without reprocessing.

The two main types of fuel cycle are the once-through thermal neutron reactor cycle and the thermal neutron reactor cycle with reprocessing. In the "once-through" thermal reactor cycle the spent fuel is not reprocessed but kept in storage until it is eventually disposed of as waste. In the thermal reactor cycle with reprocessing, the spent fuel is reprocessed and uranium and plutonium are separated from the fission products. Either the uranium, the plutonium. or both can be recycled in new fuel elements.

There are a number of thermal reactor types currently in use for electricity generation. The dominant one w worldwide is the pressurized water reactor (PWR). It has therefore been selected, both with and without reprocessing, as the nuclear reference case for this comparison. Although the other types of reactors produce wastes which are different in some details to those from a PWR, it is considered that the PWR is sufficiently representative to be the reference case.

Waste Arisings

Waste arises at each step of the fuel cycles: mining, fuel fabrication or preparation, power production, and decommissioning.

Electricity generation from nuclear fuel produces substantially different wastes both in quantity and type to those which arise from electricity generation using fossil fuel. The waste arisings from the operation of nuclear power plants are in the form of relatively small volumes of radioactive material. In contrast, fossil-fuelled plants burn large quantities of fuel and the operational waste arisings include large amounts of combustion products. Both types of power plants produce wastes in gaseous, liquid, and solid forms.

It is not widely appreciated that the combustion of coal releases quantities of radiation to the environment that are similar (in terms of its potential biological consequences) in magnitude to the routine releases from the nuclear industry for comparable electrical output. Natural gas production and usage also release radioactive radon to the atmosphere.

Fossil fuel wastes. Most of the wastes from fossil fuel cycles arise during power production, although for coal substantial amounts of solid wastes are also produced during mining and fuel preparation.

Combustion of fossil fuels produces carbon dioxide. Compared with coal, burning natural gas produces somewhat more than half the CO_2 on a per unit energy content basis. Coal combustion also produces oxides of sulphur (SO_2 and SO_3) while gas combustion products have almost no sulphur compounds. The combustion of coal and gas also produces oxides of nitrogen (NO_x).

Particulate emissions (ash) also occur with the combustion of coal. Part of the ash, about 10%, remains in the boiler and is removed; this is called bottom ash. Most of the ash, however, appears as a very fine particulate material in the flue gas; this is known as fly ash.

For fossil facilities, decommissioning will likely occur soon after the end of a plant's operating life. Decommissioning wastes will generally be those associated with demolition, and would not nose special residual hazards.

Arisings of conditioned radioactive wastes

Step	Waste category	Unit	Range		
			Low	Reference	High
Mining and milling	LLW	m³/a	20 000	40 000	60 000
Conversion and LLW m3 /a enrichment	LLW	m³/a	20	20	20
Fabrication	LLW	m³/a	20	30	30
Power plant operation	LLW	m³/a	100	130	200
	ILW	m³/a	50	30	100
Reprocessing	C1	m³/a	3.5	4	4
	C2	m³/a	20	22	25
	ILW	m³/a	50	75	100
	LLW	m³/a	470	580	690
Spent fuel (unconditioned and once-through)		t/a	25	30	35

Nuclear fuel cycle considered for assessment

```
                    ┌──────────────────┐
                    │ Mining and milling│
                    │     of ore        │
                    └──────────────────┘          uranyl  nitrate
            yellow cake      │          ◄───────────────────┐
                             ▼                               │
                    ┌──────────────────┐                     │
                    │   Enrichment     │                     │
                    │      and         │                     │
                    │   conversion     │        PuO₂         │
                    └──────────────────┘        ◄──────┐     │
            UO₂              │                          │     │
                             ▼                          │     │
                    ┌──────────────────┐                │     │
                    │  Fuel fabrication│                │     │
                    └──────────────────┘                │     │
            fuel assembly    │                          │     │
                             ▼                          │     │
                    ┌──────────────────┐                │     │
                    │       PWR        │                │     │
                    └──────────────────┘                │     │
            spent fuel       │                          │     │
                             ▼                          │     │
                    ┌──────────────────┐                │     │
                    │ Spent fuel storage│               │     │
                    └──────────────────┘                │     │
      ("once through" option)    (reprocessing option)  │     │
            ▼                          ▼                 │     │
    ┌──────────────┐           ┌──────────────┐          │     │
    │  Spent fuel  │           │ Reprocessing │──────────┘     │
    │ conditioning │           └──────────────┘                │
    └──────────────┘             conditioned                   │
      spent fuel │                  waste │                    │
            ▼                          ▼                        │
    ┌──────────────┐           ┌──────────────┐                │
    │  Spent fuel  │           │Reprocessing waste│            │
    │storage and   │           │storage and       │            │
    │  disposal    │           │  disposal        │            │
    └──────────────┘           └──────────────┘                │
```

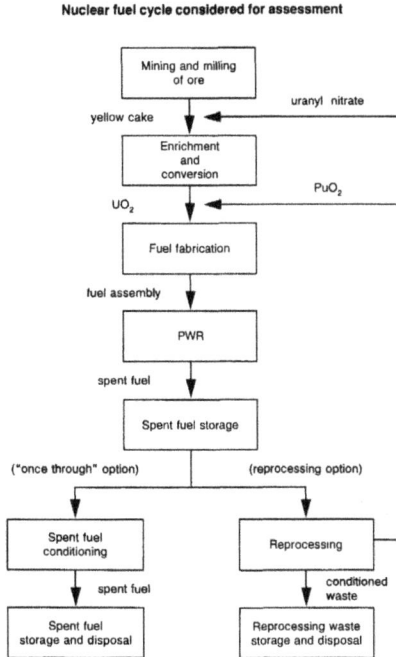

Coal Waste Streams after Waste Treatment

	Waste streams (g/kWh)
NO_x	0 25
So_2	0 32
Fly ash to air	0.07
Fly ash collected	3 02
Gypsum	2 10

Nuclear wastes. As in fossil fuel cycles, wastes occur at each stage of the nuclear fuel cycle.

Mining wastes consist mainly of mine waters and waste rock piles. While uranium mill tailings are generally similar to mining wastes, they contain nearly all of the naturally occurring radioactive daughters from the decay of uranium.

The conversion and enrichment processes produce solid and gaseous wastes which contain some airborne uranium. In addition, enrichment plants produce large volumes of depleted uranium which is considered as a waste for purposes of this assessment.

Depending on whether reprocessing is employed, fuel fabrication wastes are in the form of various solid and liquid streams contaminated with uranium and/or plutonium.

Radioactivity occurs in various liquid waste streams of the power plant. In addition, small quantities of gaseous waste are generated during reactor operation. Reactor operations also give rise to a range of solid wastes in the form of contaminated or activated components.

The radioactive content of reprocessing wastes consists largely of the fission and activation products and minor actinides that are fed into the reprocessing plant as part of the spent fuel. They occur as a variety of solid and liquid waste streams.

The radioactive nature of some of the components of nuclear fuel cycle facilities requires expensive remote handling techniques to be employed during decommissioning. The cost and need for such an approach can be reduced by delaying the work and allowing decay of radioactivity. For nuclear power plants, deferred decommissioning is a strategy commonly employed throughout the world and has been selected as the reference for the purpose of this assessment. Most of the radioactive waste from decommissioning nuclear fuel cycle facilities is low-level (LLW) solid waste. Small components of intermediate level (ILW) and high-level (HLW) or transuranic waste are associated with reprocessing of spent fuel and the fabrication of mixed-oxide fuel.

Waste Management

Fossil wastes: The largest solid waste management problem faced by the coal cycle is that related to mine spoils. These pose a significant disposal problem. One possible option includes backfilling in the mines from which they came.

The flue gas treatment in a coal plant consists of three steps: NO_x removal, SO_2 reduction, and particulate reduction. For natural gas, the only significant flue gas waste management problem currently capable of resolution is that associated with nitrogen oxides.

Modest NO_x reduction is achieved by modification of the combustion process. However the most effective process for NOX removal is selective catalytic reduction (SCR) which uses ammonia and a catalyst to reduce the NO_x to nitrogen and water. Typical SCR reduction rates are around 80%. Used catalytic material is the only waste that requires disposal as a result of this process. However, the main component can be returned to the supplier for reuse.

Flue gas desulphurization (FGD) processes use alkaline materials to absorb and remove the sulphur dioxide from the flue gas. FGD processes tend to be large and expensive since very large volumes of flue gas containing very low concentrations of SO_2 must be treated. Large quantities of product (gypsum) result from the reaction of the sulphur dioxide and the reagent. Some of this can be treated to produce a wallboard quality gypsum and the rest is disposed of as landfill. Typical FGD removal efficiencies are 95%.

The reduction of particulate in the flue gas is usually achieved using electrostatic precipitators (ESP) which typically have a removal efficiency of about 95%. Some of the removed fly and bottom ash can be utilized in the cement and road building industries, the remainder requires disposal as landfill.

At present, there is no cost effective technology which will reduce CO_2 emissions and no attempts have been made on removal of radionuclides from gas effluents. Various waste streams are discharged into the environment after waste treatment.

Nuclear wastes: A number of techniques are currently used in the management of radioactive wastes. These range from direct discharge to the environment (dispersal) to sophisticated techniques for immobilization of the radionuclides and their disposal in carefully designed and constructed disposal facilities.

Mining and milling waste. All wastes arising from the milling of uranium ores are treated before any release takes place. The disposal of mill tailings is usually done on-site, often by covering the tailings to reduce radioactive dispersion.

Liquid wastes. Liquid waste treatment forms a significant part of the waste management scheme at most nuclear facilities. The waste management option depends on the characteristics of the waste and the quantity being produced. Small quantities of aqueous wastes containing short-lived radionuclides may be discharged into the environment. Liquid wastes containing large salt concentrations can be evaporated with the radioactive material being retained in the concentrate or being chemically precipitated to produce a sludge with suitable properties for further treatment. Some liquid wastes can be absorbed on solid matrices, again as a precursor to further treatment of the solid. Incineration is also sometimes used for volume reduction of active oils and combustible solvents. LLW and ILW concentrates are encapsulated in cement or bitumen matrices, and then packaged in suitable containers.

Liquid HLW from a reprocessing facility contains almost all of the fission products produced in the fuel. Currently such HLW is converted into glass using a vitrification process and the molten glass is cast into stainless steel 1containers prior to disposal in a suitable deep repository. (These high-level heat emitting wastes are classified as Cl in this assessment.)

Gaseous wastes. Radioactive gaseous wastes are usually discharged in the atmosphere in accordance with the appropriate regulatory requirement. Before discharge, the gaseous wastes are treated, if necessary, to ensure that the regulatory limits on the discharges are not exceeded.

Coal: Conventional Steam Cycle

Fossil fuel waste management cost estimates

0.20%

14%   8%

30%

48%

Gas: Combined Cycle with SCR

1%

99%

SCR = selective catalytic reduction

Legend:
- Front End
- SO$_2$ Removal
- NO$_x$ Removal
- Particulate Removal
- Decommissioning

Comparison of levelized costs: Fossil fuel waste management

Mills/kWh

30.00
25.00
20.00
15.00
10.00
5.00
0.00

Low   Reference   High

Cost Scenario

- Conventional steam cycle-Coal
- Combined cycle-Gas (with SCR)

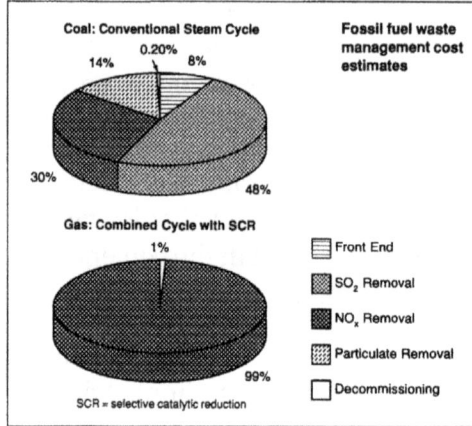

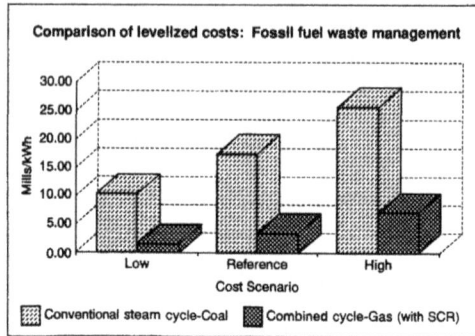

Solid wastes: Apart from already mentioned vitrified reprocessing wastes, solid wastes also include cladding hulls and fuel assembly hardware (classified as C2), filters, used equipment, resins and sludges, scrubber solids, and general trash. All of the waste, except that with very low activity levels, will need some treatment and conditioning.

Treatment and conditioning operations include volume reduction, conversion of the waste to more stable forms, and packaging. The various stages of waste management for the PWR cycle considered here produce different volumes of conditioned solid wastes.

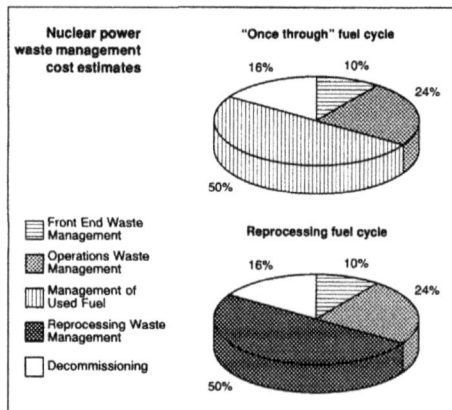

Nuclear power waste management cost estimates

"Once through" fuel cycle

16%   10%

24%

50%

Legend:
- Front End Waste Management
- Operations Waste Management
- Management of Used Fuel
- Reprocessing Waste Management
- Decommissioning

Reprocessing fuel cycle

16%   10%

24%

50%

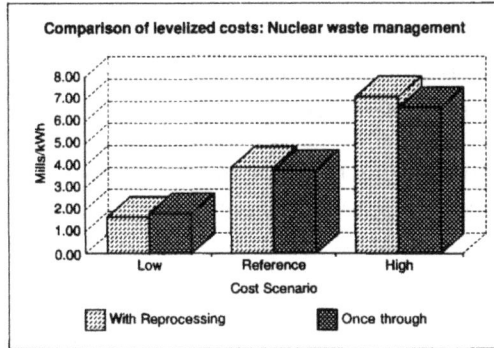

Comparison of levelized costs: Nuclear waste management

Disposal in a suitable facility, which may be deep geologic or near surface, contributes to limiting any transport of radionuclides into the environment to acceptable levels. For the once through cycle, spent fuel is stored for a period of years, probably several decades to allow the radioactivity and associated heat load to decay before disposal.

Methodology for Cost Assessment

The data for each of the cost components of waste management have been obtained from a survey of existing estimates. In order to provide a basis for a meaningful comparison of costs, the raw data have been adjusted where necessary and applied to the reference cases. Finally, all cost estimates have been converted to a common basis of levelized unit energy (LUC) costs expressed in US dollars as of 1 July 1991, per kWh. The LUC is defined such that the present value of the cost stream equals the present value of the single value levelized cost times the number of units (kWh) in each time-frame. In order to put waste management costs on a common basis for comparison purposes, it is necessary to convert all cost flows to a common value by the procedure of discounting. This is widely accepted in economic assessments as a procedure which facilitates the comparison of investment options having distinct cash flows spread out in time.

The major criticism of applying the discounting technique to the assessment of the cost of nuclear power is its application to significant cost streams long after the production of electricity from the nuclear generating station ceases. This criticism relates to intergenerational equity — that is, the extent to which electricity customers pay the full costs of serving them and the extent to which future generations bear costs from which they receive no benefit.

In order to recognize this concern, the reference cases are based on a 5% real discount rate to the end of power plant life followed by a zero discount rate thereafter. A 5% real discount rate is favoured by many countries of the Organization for Economic Co-operation and Development (OECD). In addition, the results have been tested for sensitivity to different factors: the discount rate, the capacity factor, and the service life of the power plant.

Cost Data

Fossil fuel cycle: For both fuel cycles the levelized waste management costs cover a range of about 0.5 to 2.0 times the reference cases.

When looking at the relative proportions of these costs, the control of SO_2 alone contributes about 48% of the costs in the conventional steam cycle coal plant. For the combined cycle, 99% of the waste management costs are comprised of the decommissioning cost.

Fossil fuel waste management costs are in the range of close to zero to about 25 mills per KWh (a mill is one-thousandth of a US dollar). The costs are expected to remain in this range with typical variations in capacity factor discount rate or service life. The low end of the range corresponds to gas-fired generation and the high end to coal-fired generation. At these levels, waste management costs represent a low to moderate fraction of the overall cost of base load electricity generation from fossil fuels. Total levelized costs of fossil-based electricity generally fall in the range of 40-60 mills per kWh.

Nuclear fuel cycle: The levelized cost of waste management for the two nuclear fuel cycles assessed are similar.

For both cycles, waste management at the front end of the cycle leads to about 10% of the total waste management cost. Of this about one-third is due to the management of depleted uranium as a waste. The management of wastes from power plant operation accounts for about 24% of the costs and 15% is due to power plant decommissioning. The remaining 50% of costs is associated with the back end of the fuel cycle.

Nuclear waste management costs are in the range of 1.6 mills/kWh to 7.1 mill/kWh. As in the case of fossil waste management, such costs represent a low to moderate fraction of the cost of electricity generated. The waste management costs may be compared to the cost of nuclear powered electricity, which is 30 to 50 mills/kWh.

Comparison: The waste management costs for the nuclear cases lie between those of the two fossil cases. They are closest to the costs for gas-fuelled combined cycle, which represents the lower end of the fossil range. The coal-fuelled option, representing the top end of the fossil range, has waste management costs which are about a factor of four above those of the nuclear cases.

While both the nuclear cases show a range between the high and low values covering a factor of four, the variability in fossil cost estimates only cover a factor of two or less. This difference in variability can in part be attributed to the fact that the fossil costs are based on established technology, while the nuclear costs include a substantial contribution from waste management activities which have yet to be firmly established. Even though flue gas treatment is a relatively new field, several plants are in operation and the cost estimates are firmer than those for some nuclear waste management techniques, such as decommissioning and deep repositories. In light of this, there is greater un-certainty associated with nuclear waste management costs than with those fossil waste management activities considered in this assessment. Some of the difference in variability between fossil and nuclear waste management cost estimates is also due to the effect of differences between local conditions, including regulatory requirements.

Possible Future Changes

For nuclear generation a major shift in waste management practices or expectations is not foreseen. Nonetheless the future holds some possibilities that could influence waste management costs. These include attempts to increase fuel burn-up, better housekeeping, and more effective and advanced waste treatment techniques, such as super compaction, biodegradation, incineration, and plasma torch burning. All of these developments hold the promise of reducing nuclear waste management costs. The future will also bring the development of deep repositories and much greater experience with decommissioning. While these bring with them the risk that costs might turn out to be higher than expected, they will also significantly reduce the uncertainty with respect to nuclear waste management costs.

In the case of fossil waste management costs, one of the major developments is expected to be the more wide spread use of clean coal technologies. This will result in reduced environmental impacts and waste management costs through a technology that better integrates emission control within the power generation process itself. A further possible development related to fossil-fuelled generation is regulation with respect to CO_2. This could involve the development of technological solutions such as the disposal of CO_2 in empty gas fields at the bottom of the ocean, or the introduction of carbon taxes —both of which could significantly increase fossil waste management costs.

For both nuclear and fossil cases, there is also the possibility that existing waste management regulations will be further tightened. This would include the possibility that residual environmental costs would have to be internalized by electric utilities. Such changes would bring with them increased costs.

References

- Pulp-mill-wastewater-characteristics-and-treatment, biological-wastewater-treatment-and-resource-recovery: intechopen.com, Retrieved 15 July 2018

- Recycling-rubber: seas.columbia.edu, Retrieved 11 April 2018

- Safe-disposal-treated-timber, household-building-and-renovation: epa.nsw.gov.au, Retrieved 21 June 2018

- Waste-management, how-to-treat-waste-in-bakery-industry-waste-management-5220: environmentalpollution.in, Retrieved 27 May 2018

- Bulletin, magazines, bull35-4: iaea.org, Retrieved 14 April 2018

Permissions

All chapters in this book are published with permission under the Creative Commons Attribution Share Alike License or equivalent. Every chapter published in this book has been scrutinized by our experts. Their significance has been extensively debated. The topics covered herein carry significant information for a comprehensive understanding. They may even be implemented as practical applications or may be referred to as a beginning point for further studies.

We would like to thank the editorial team for lending their expertise to make the book truly unique. They have played a crucial role in the development of this book. Without their invaluable contributions this book wouldn't have been possible. They have made vital efforts to compile up to date information on the varied aspects of this subject to make this book a valuable addition to the collection of many professionals and students.

This book was conceptualized with the vision of imparting up-to-date and integrated information in this field. To ensure the same, a matchless editorial board was set up. Every individual on the board went through rigorous rounds of assessment to prove their worth. After which they invested a large part of their time researching and compiling the most relevant data for our readers.

The editorial board has been involved in producing this book since its inception. They have spent rigorous hours researching and exploring the diverse topics which have resulted in the successful publishing of this book. They have passed on their knowledge of decades through this book. To expedite this challenging task, the publisher supported the team at every step. A small team of assistant editors was also appointed to further simplify the editing procedure and attain best results for the readers.

Apart from the editorial board, the designing team has also invested a significant amount of their time in understanding the subject and creating the most relevant covers. They scrutinized every image to scout for the most suitable representation of the subject and create an appropriate cover for the book.

The publishing team has been an ardent support to the editorial, designing and production team. Their endless efforts to recruit the best for this project, has resulted in the accomplishment of this book. They are a veteran in the field of academics and their pool of knowledge is as vast as their experience in printing. Their expertise and guidance has proved useful at every step. Their uncompromising quality standards have made this book an exceptional effort. Their encouragement from time to time has been an inspiration for everyone.

The publisher and the editorial board hope that this book will prove to be a valuable piece of knowledge for students, practitioners and scholars across the globe.

Index

A

Aeration, 126-127, 144, 156, 170-173, 190, 203-205

Aerobic Wastewater Treatment, 190

Air Sparging, 105-106

Anaerobic Filter, 194-195, 197

Anaerobic Fluidized Bed Reactor, 194-195

Autoclaving, 36-39, 64, 85-86

B

Bakery Waste, 197, 199-202

Bioaugmentation, 111, 157

Biohazardous Waste, 35-36, 47, 63-64, 70-71

Biological Treatment, 9-10, 91, 127-128, 132, 140, 167-172, 189-191, 197, 199, 201-203

Biological Waste, 63-64, 66, 203

Bioremediation, 92, 99-100, 110, 112, 128

Bioventing, 108-109

Bleaching Effluent, 166, 169

C

Centrifugation Method, 95-96

Chemical Recycling, 115, 118-119

Chemical Treatment, 9, 65, 140, 190-191, 200

Chemical Waste, 4-5, 46, 49, 52-54, 62-63

Coagulation-flocculation, 185, 201

Concrete Pit, 141, 144, 146

Contaminated Soil, 103, 105, 141, 157

Corrosivity, 23-26

Cytotoxic, 51-52, 74, 86

D

Disposal Method, 72, 98

Dissolved Air Flotation, 91, 190, 199, 201-202

E

Evaporation Bed, 145

F

Fossil Fuel Cycle, 206, 214

G

Genotoxic Waste, 49, 51-52

H

Health Care Waste, 59

Household Hazardous Waste, 12-15, 21

Hydrolysis, 147, 163, 167, 201

I

Ignitability, 6, 17, 23-24

Incineration, 4, 9-10, 34, 36, 41, 63, 65, 67, 80-83, 86, 92, 98, 102, 105, 108, 116, 119, 140, 147-149, 179, 182, 211, 215

Incinerator, 35-36, 63-64, 75, 80-81, 83, 85-86, 98, 148

Infectious Waste, 39-44, 48-50, 53, 61-63, 67-69, 75, 77, 80

K

Kpeg, 147, 154

Kraft Pulping, 160-162, 164-165, 170

L

Land Cultivation, 140-142

Landfilling, 116, 182

M

Mechanical Pulping, 159, 161, 163-164

Medical Waste, 32, 35, 39, 45-46, 72-75, 77-81, 83, 85-86

N

Non-hazardous Biological Waste, 63, 66

Nuclear Waste Management, 214-215

O

Oily Sludge Source, 91

Open Burning, 147, 149

Ozonation, 147, 149

P

Pathological Waste, 49-50, 65-66, 75, 81

Pesticide Waste, 130, 138, 142-143, 145, 149

Physical Treatment, 8, 140, 190

Phytoremediation, 112, 147, 156-157

Plastic Lined Pit, 142, 144

Potentially Infectious Medical Waste, 45-46

R
Radioactive Substance, 57
Radioactive Waste, 41, 46, 49, 56, 58-60, 73, 87, 205, 210
Radioactivity, 56-57, 61, 87, 209-210, 213

S
Sedimentation, 126-127, 170-171, 190, 199, 201-202, 204
Septic System, 132
Sharps Container, 38, 70, 77
Sharps Waste, 38, 47, 68-73
Silk Waste, 113, 121
Silver Bearing Waste, 131
Silver Recovery, 133-135
Sludge Treatment, 92, 94, 96, 100, 171, 204
Soft Drink Waste, 183, 196
Soil Pit, 141-143
Solid Waste, 3, 5, 21, 29, 34, 59, 76, 102, 115, 199, 210
Solidification, 98-99, 102-103, 107-108

Stabilization, 98-99, 107-108
Sulphite Pulping, 159, 161-162
Synthetic Rubber, 175, 177-178

T
Textile Recycling, 113, 115, 119, 122-124, 129
Textile Wastewater
Treatment, 113, 126-127
Thermal Process, 85
Thermal Treatment, 9, 64, 103, 146
Toxicity, 6, 23-24, 29-30, 34, 105, 128, 140, 142, 157, 168, 170, 172-173, 199

U
Ultrasonic Irradiation, 97, 99

V
Vulcanisation, 175-177, 180

W
Waste Disposal, 4-5, 7, 10, 12, 48, 66, 73, 201
Waste Minimization, 7, 88